Multimodal Perception and Secure State Estimation for Robotic Mobility Platforms

IEEE Press
445 Hoes Lane
Piscataway, NJ 08854

Multimodal Perception and Secure State Estimation for Robotic Mobility Platforms

Xinghua Liu
Xi'an University of Technology
China

Rui Jiang
National University of Singapore
Singapore

Badong Chen
Xi'an Jiaotong University
China

Shuzhi Sam Ge
National University of Singapore
Singapore

IEEE PRESS
WILEY

Published by John Wiley & Sons, Inc., Hoboken, New Jersey.

Published simultaneously in Canada.

Library of Congress Cataloging-in-Publication Data applied for

Hardback: 9781119876014

Cover Design: Wiley
Cover Image: © Darq/Shutterstock.com

Set in 9.5/12.5pt STIXTwoText by Straive, Chennai, India

To

Yu Xia, Siyu
and
our loved ones

Contents

About the Authors

Xinghua Liu received the B.Sc. from Jilin University, Changchun, China, in 2009; and the Ph.D. degree in Automation from University of Science and Technology of China, Hefei, in 2014. From 2014 to 2015, he was invited as a Visiting Fellow at RMIT University in Melbourne, Australia. From 2015 to 2018, he was a Research Fellow at the School of Electrical and Electronic Engineering in Nanyang Technological University, Singapore. Dr. Liu has joined Xi'an University of Technology as a Professor since September 2018. His research interest includes state estimation and control, intelligent systems, autonomous vehicles, cyber-physical systems, robotic systems, etc.

Rui Jiang received the B.Eng. degree in Measurement, Control technique and Instruments from Harbin Institute of Technology, Harbin, China, in 2014; and the Ph.D. degree in Control, Intelligent systems & robotics from National University of Singapore, Singapore, in 2019. Dr. Jiang is an Adjunct Lecturer with the Department of Electrical and Computer Engineering, National University of Singapore. His research interests includes intelligent sensing and perception for robotic systems.

Badong Chen received the B.S. and M.S. degrees in control theory and engineering from Chongqing University, Chongqing, China, in 1997 and 2003, respectively, and the Ph.D. degree in Computer Science and Technology from Tsinghua University, Beijing, China, in 2008. He was a Postdoctoral Researcher with Tsinghua University from 2008 to 2010, and a Postdoctoral Associate with the University of Florida Computational NeuroEngineering Laboratory from 2010 to 2012. He is currently a Professor with the Institute of Artificial Intelligence and Robotics, Xi'an Jiaotong University, Xi'an, China. He has authored or coauthored two books, four chapters, and more than 200 papers in various journals and conference proceedings. His research interests include signal processing, machine learning, and their applications to neural engineering and robotics. Dr. Chen is a Member of the Machine Learning for Signal Processing Technical Committee of the IEEE Signal Processing Society and the Cognitive and Developmental Systems Technical Committee of the IEEE Computational Intelligence Society. He is an Associate Editor for the *IEEE Transactions on Cognitive and Developmental Systems*, *IEEE Transactions on Neural Networks and Learning Systems*, and the *Journal of the Franklin Institute*, and has been on the Editorial Board of Entropy.

Shuzhi Sam Ge received his Ph.D. degree and Diploma of Imperial College (DIC) from the Imperial College London, London, U.K., in 1993, and B.Sc. degree from the Beijing University of Aeronautics and Astronautics (BUAA), China, in 1986. He is a Professor with the Department of Electrical and Computer Engineering, PI member of Institute for Functional Intelligent Materials, the National University of Singapore, Singapore, and the Founding Honorary Director of Institute for Future (IFF), Qingdao University, Qingdao, China. He serves as the Founding Editor-in-Chief, International Journal of Social Robotics, Springer Nature, and book editor for Automation and Control Engineering of Taylor & Francis/CRC Press. He has served/been serving as an Associate Editor for a number of flagship journals including IEEE TAC, IEEE TCST, IEEE TNN, IEEE Transaction on SMC-Systems, Automatica, and CAAI Transactions on Intelligence Technology. At Asian Control Association, he serves as President Elec, 2022–2024. At IEEE Control Systems Society, he served as Vice President for Technical Activities, 2009–2010, Vice President of Membership Activities, 2011–2012, Member of Board of Governors of IEEE Control Systems Society, 2007–2009. He is a Clarivate Analytics high-cited scientist in 2016–2021. He is a Fellow of IEEE, IFAC, IET, and SAEng. His current research interests include robotics, intelligent systems, artificial intelligence, and smart materials.

Preface

Cyber-physical systems (CPSs), which refer to the embedding of widespread sensing, networking, computation, and control into physical spaces, play a crucial role in many areas today. As an important application area of CPSs, autonomous vehicles have emerged over the past few years as a subdiscipline of control theory in which the flow of information in a system takes place across a communication network. Unlike traditional control systems, where computation and communications are usually ignored, the approaches that have been developed for autonomous vehicle systems explicitly take into account various aspects of the communication channels that interconnect different parts of the overall system and the nature of the distributed computation that follows from this structure. This leads to a new set of tools and techniques for analyzing and designing autonomous vehicles that builds on the rich frameworks of communication theory, computer science, and control and estimation theory.

This book is intended to be employed by researchers and graduate-level students interested in sensor fusion and secure state estimation in mobile robots or autonomous vehicles. The book is aimed to be self-contained and to give interested readers insights into modeling, analysis, and applications of techniques of CPS-based autonomous vehicles. We also provide pointers to the literature for further reading on each explored topic. Moreover, numerous illustrative figures and step-by-step examples help readers understand the main ideas and implementation details.

This book is organized in two parts: multimodal perception for vehicle pose estimation and secure state estimation for mobile robots. The first part discusses different sensor configurations and introduces new sensor fusion algorithms and frameworks to minimize pose estimation errors. Those concepts and methods could be used in current state-of-the-art autonomous vehicles, and extensive experimental results are provided to verify algorithm performance on real robotic platforms. In the second part, we deal with the problem of secure pose estimation for mobile robots under many different types of attacks (possible sensor attacks).

A filter-based secure dynamic pose estimation approach is presented such that the vehicle pose can be resilient under randomly occurring deception attacks. Based on the established heading measuring model, we can decompose the optimal Kalman estimate into a linear combination of local state estimates, and a convex optimization-based approach is introduced to combine the local estimate into a more secure estimate under the (p, m) sparse attack.

In particular, Chapter 1 introduces multimodal pose estimation and secure estimation for vehicle navigation, as well as the organization of the entire book. Chapter 2 presents an optimization-based sensor fusion framework where absolute heading measurements are used in vehicle pose estimation. Two road-constrained pose estimation methods are introduced in Chapter 3 and Chapter 4 to reduce error and drift for long travel distances. Chapter 5 presents a unified framework to combine heading reference and map assistance, and the framework can be applied to other types of measurements and constraints in vehicle pose estimation. The core material on secure dynamic pose estimation for autonomous vehicles is presented in Chapter 6, where an upper bound for the estimation error covariance is guaranteed to establish stable estimates of the pose states. Chapter 7 presents a pose estimation approach for ground vehicles under randomly occurring deception attacks, and an unscented Kalman filter-based secure pose estimator is then proposed to generate a stable estimate of the vehicle pose states. Chapters 8 and 9 then go on to consider secure dynamic state estimation under (p, m) sparse attacks. We prove that we can decompose the optimal Kalman estimate as a weighted sum of local state estimates, and a convex optimization-based approach is introduced to combine the local estimate into a more secure state estimate. It is shown that the proposed secure estimator coincides with the Kalman estimator with a certain probability when there is no attack and can be stable when p elements of the model state are compromised. In each of these chapters on the core material, we have attempted to present a unified view of many of the most recent and relevant results in secure state estimation to establish a foundation on which more specialized results of interest to specific groups can be covered.

We would like to thank Prof. Han Wang, Prof. Yilin Mo, Prof. Emanuele Garone, Prof. Tong Heng Lee, Dr. Shuai Yang, Dr. Hui Zhou, Dr. Handuo Zhang, and Dr. Xiaomei Liu for their collaboration in this research area in recent years. We thank Prof. Xinde Li, Dr. Mien Van, and Dr. Yuanzhe Wang for carefully reviewing the book and for the continuous support given to us in the entire publication process. In addition, we would like to thank the National Key Research and Development Program of China (No. 2020YFB1313600), the National Natural Science Foundation of China, the Natural Science Foundation of

Shaanxi Province, the Key Laboratory Project of Shaanxi Provincial Department of Education, and Shaanxi Youth Science and Technology New Star Project for their financial support of our research work.

Xinghua Liu
Xi'an University of Technology, China

Rui Jiang
National University of Singapore, Singapore

Badong Chen
Xi'an Jiaotong University, China

Shuzhi Sam Ge
National University of Singapore, Singapore

1

Introduction

1.1 Background and Motivation

With the rapid development of sensor technologies, and due to increased density in integrated circuits predicted by Moore's law, the autonomous vehicle has become a fruitful area blending robotics, automation, computer vision, and intelligent transportation technologies. It has been reported that traditional automobile companies and startups plan to get their autonomous driving systems ready in the 2020s [Ross, 2017].

The US Department of Transportation's National Highway Traffic Safety Administration (NHTSA) defined five levels of autonomous driving, from manual driving (level 0), to driver assistance (level 1), to fully autonomous driving (level 5) (https://www.sae.org/standards/content/j3016_202104). As an inspiring example, the Audi A8, launched in 2017, is claimed to be "the world's first production automobile conditional automated at level 3," according to Audi AG. Nevertheless, some pessimistic voices have emerged, claiming that fully autonomous cars will not be developed as quickly as expected or are even unlikely. One of the pacesetters in fully autonomous driving technologies, Waymo LLC, has received resident complaints due to conflicts in driving behaviors between humans and autonomous vehicles.

Although it is still a long way to level 5 autonomy, there is high demand for the development of autonomous vehicles so that tasks related to logistics, environmental cleanup, public security, and much more can be automated. Among all the functional blocks in autonomous vehicles, the navigation system plays an irreplaceable role since the vehicle needs to be literally "in motion" for any particular task. Multimodal perception and state estimation are two coadjutant modules for vehicle navigation. There have been extensive research outcomes on these two

Multimodal Perception and Secure State Estimation for Robotic Mobility Platforms, First Edition.
Xinghua Liu, Rui Jiang, Badong Chen, and Shuzhi Sam Ge.

topics in autonomous vehicle navigation, but a few challenges still exist, motivated by which the in-depth studies in this book have been carried out:

- A modern pose estimation system contains multiple sensors to achieve accuracy and robustness. Appropriate sensor configurations, which combine the advantages of each sensor to benefit the whole estimation system, are distinct depending on the specific applications and requirements. Based on a particular sensor configuration, new theories and ideas are required for multi-sensor pose estimation, where states, measurements, and constraints are represented in a unified fusion framework.
- Due to the stealthiness of attacks, system operators usually cannot discover attacks in time, which may lead to severe economic damage and even the loss of human lives. Such incidents indicate that enhancing the security of the system is an urgent issue. Researchers have studied how we can securely estimate the state of a dynamical system from the controller's point of view based on a set of noisy and maliciously corrupted sensor measurements. In particular, researchers have focused on linear dynamical systems and have tried to understand how the system dynamics can be leveraged for security guarantees.

This book discusses the pose estimation problem for robotic mobility platforms using information from multiple sensors. The first part discusses different sensor configurations and introduces new sensor fusion algorithms and frameworks to minimize pose estimation errors. Those concepts and methods are extensively used in current state-of-the-art autonomous vehicles, and extensive experimental results have been provided to verify the algorithm performance on real robotic platforms. The second part focuses on the secure estimation problem in multi-sensor fusion, where attacks are considered and explicitly modeled in algorithm design. As this is a new topic that is at the primary stage of research, theoretical analysis and simulation results are shown in the related chapters.

1.2 Multimodal Pose Estimation for Vehicle Navigation

1.2.1 Multi-Senor Pose Estimation

Multi-sensor fusion is a typical solution where system dynamics, measurements, and constraints are fused consistently to increase estimation performance in terms of accuracy and robustness [Borges and Aldon, 2002, Ye et al., 2015, Teixeira et al., 2018]. Essentially, pose estimation can be considered as state estimation within a state space with a problem-dependent topological structure. Let us assume the following discrete state equation and output equation:

$$\mathbf{x}_k = \mathbf{f}(\mathbf{x}_{k-1}, \mathbf{u}_{k-1}) + \mathbf{w}_{k-1} \tag{1.1}$$

$$\mathbf{z}_k = \mathbf{h}(\mathbf{x}_k) + \mathbf{v}_k \tag{1.2}$$

where $\mathbf{x}_k$, $\mathbf{u}_{k-1}$, $\mathbf{z}_k$ denote the state, control input, and measurement, respectively; $\mathbf{f}(\cdot)$ and $\mathbf{h}(\cdot)$ are the C^∞ state equation and output equation; and $\mathbf{w}_{k-1}$ and $\mathbf{v}_k$ represent process and measurement noise.

Filtering and optimization are two frequently used data fusion frameworks for pose estimation. Filtering approaches propagate state vectors with their joint probability distributions along with time. The Kalman filter models the state and noise as Gaussian, which is not suitable for non-Gaussian or multimodal distributions. The particle filter and its variants [Van Der Merwe et al., 2001, Nummiaro et al., 2003] have been proposed to deal with non-linear and non-Gaussian systems, and the computation load of updating particle states proliferates with the sample number. The optimization-based approaches retain historical measurement and estimation as a graph such that they can be used for bundle adjustment or simultaneous localization and mapping (SLAM) [Grisetti et al., 2010]. The two commonly used frameworks are elaborated here.

Filtering-Based Approaches As shown in related work [Janabi-Sharifi and Marey, 2010, Koval et al., 2015, Bloesch et al., 2017], filters provide a probabilistic solution on pose estimation, which can be divided into two steps. First, the "prediction" step predicts states without current measurement, according to the state equation

$$\mathbb{P}(\mathbf{x}_k|\mathbf{z}_{1:k-1}) = \int \mathbb{P}(\mathbf{x}_k|\mathbf{x}_{k-1})\mathbb{P}(\mathbf{x}_{k-1}|\mathbf{z}_{1:k-1})\mathrm{d}\mathbf{x}_{k-1} \tag{1.3}$$

where $\mathbb{P}(\cdot|\cdot)$ denotes the conditional distribution, and specifically $\mathbb{P}(\mathbf{x}_k|\mathbf{x}_{k-1})$ is obtained from (1.1). Then, the probability distribution of the update can be obtained in the "correction" step, based on the output equation

$$\mathbb{P}(\mathbf{x}_k|\mathbf{z}_{1:k}) = \frac{\mathbb{P}(\mathbf{z}_k|\mathbf{x}_k)\mathbb{P}(\mathbf{x}_k|\mathbf{z}_{1:k-1})}{\mathbb{P}(\mathbf{z}_k|\mathbf{z}_{1:k-1})} \tag{1.4}$$

where $\mathbb{P}(\mathbf{z}_k|\mathbf{x}_k)$ is obtained from (1.2), and the constant denominator is

$$\mathbb{P}(\mathbf{z}_k|\mathbf{z}_{1:k-1}) = \int \mathbb{P}(\mathbf{z}_k|\mathbf{x}_k)\mathbb{P}(\mathbf{x}_k|\mathbf{z}_{1:k-1})\mathrm{d}\mathbf{x}_k \tag{1.5}$$

Optimization-Based Approaches Instead of using the filtering-based approaches, some other research [Leutenegger et al., 2015, Huang et al., 2017, Parisotto et al., 2018, Wang et al., 2018a] aims to minimize the user-defined cost function $J(\mathbf{x})$ such that

$$\mathbf{x}^* = \arg\min_{\mathbf{x}} J(\mathbf{x}) = \arg\min_{\mathbf{x}} \sum_{C} \mathbf{e}(\mathbf{z},\mathbf{h}(\mathbf{x}))^{\mathsf{T}}\boldsymbol{\Omega}\,\mathbf{e}(\mathbf{z},\mathbf{h}(\mathbf{x})) \tag{1.6}$$

where C denotes the cost items to be considered; the information matrix $\boldsymbol{\Omega}$ indicates the degree of confidence in the corresponding measurement; and the error function $\mathbf{e}(\mathbf{z},\mathbf{h}(\mathbf{x}))$ measures the difference between the ideal and actual measurement.

1.2.2 Pose Estimation with Constraints

Constraints[1] in pose estimation are helpful in increasing algorithm robustness and accuracy. For example, we may consider *motion constraints* (1.1), that limit the vehicle's pose change with time, and *road constraints*, which require the vehicle to stay on the road. Constraints in practical issues are mostly considered as soft to allow modeling errors and noise. We discuss constrained pose estimation from two perspectives.

Incorporating Constraints into Filtering Given the constraints $\mathbf{c}(\mathbf{x}_k) = \check{\mathbf{c}}(\mathbf{x}_k) + \check{\mathbf{z}}_k = \mathbf{0}$, where $\check{\mathbf{z}}_k$ is a constant vector, the augmented output equation can be obtained to incorporate the constraints into measurements [Mourikis and Roumeliotis, 2007, Simon, 2010, Boada et al., 2017, Ramezani et al., 2017, Yang et al., 2017a, Shen et al., 2017]:

$$\begin{bmatrix} \mathbf{z}_k \\ \check{\mathbf{z}}_k \end{bmatrix} = \begin{bmatrix} \mathbf{h}(\mathbf{x}_k) \\ -\check{\mathbf{c}}(\mathbf{x}_k) \end{bmatrix} + \begin{bmatrix} \mathbf{v}_k \\ \check{\mathbf{v}}_k \end{bmatrix} \tag{1.7}$$

where the covariance matrix of $\check{\mathbf{v}}_k$ indicates the confidence in the soft constraints. With a prediction that remains the same, the correction step can be achieved by applying the augmented output equation.

In addition, we may first obtain the estimate without constraints $\hat{\mathbf{x}}_k$ and then project the unconstrained estimates toward the constraint states to get the final estimate $\hat{\mathbf{x}}_k^*$

$$\hat{\mathbf{x}}_k^* = \arg\min_{\mathbf{x}} \left[\mathbf{x} \ominus \hat{\mathbf{x}}_k\right]^{\mathrm{T}} \mathbf{W} \left[\mathbf{x} \ominus \hat{\mathbf{x}}_k\right] \text{ s.t. } \mathbf{c}(\mathbf{x}_k) \approx \mathbf{0} \tag{1.8}$$

where $\ominus$ is an operator indicating the difference between states and $\mathbf{W}$ is a positive-definite weighting matrix. For linear systems under linear constraints, if $\mathbf{x} \in \mathbb{R}^n$, the ordinary vector subtraction is selected as $\ominus$, leading to analytical solutions. Numerical methods are required to generalize the projection method to non-linear systems or with non-linear constraints. For particle filters, particle weights can be adjusted to reduce the influence of estimation results that do not satisfy the constraints.

Incorporating Constraints into Optimization For hard constraints, the method of Lagrange multipliers can be used to construct the corresponding non-constrained optimization problem. For soft constraints, one naive but effective way is to add

1 Note that the constraints in this book, which are derived from physical principles or engineering assumptions, should be differentiated from the measurements, which are obtained from sensors.

the penalty functions to the cost function $J(\mathbf{x})$, such that

$$\mathbf{x}^* = \arg\min_{\mathbf{x}} \left(J(\mathbf{x}) + \sum_{C} \mathbf{c}^i(\mathbf{x})^{\mathrm{T}} \boldsymbol{\Omega}^i \mathbf{c}^i(\mathbf{x}) \right) \tag{1.9}$$

where $\mathbf{c}^i(\mathbf{x})$ denotes the i-th constraint to be considered; $\boldsymbol{\Omega}^i$ indicates the degree of confidence in the i-th constraint. Examples of related work can be found in [Estrada et al., 2005, Levinson et al., 2007, Lu et al., 2017, Hoang et al., 2017].

Besides the constraints discussed previously (so-called *state constraints* in the literature), *measurement constraints* can be seen in practice. One example would be the constant norm constraint on measurement vectors for translationally static but rotating magnetometers. Unfortunately, the current literature pays less attention to measurement constraints than state constraints. In Chapters 4 and 5, by presenting a unified representation containing state space and measurement space, both state constraints and measurement constraints are considered in the proposed geometric pose estimation framework.

1.2.3 Research Focus in Multimodal Pose Estimation

In the first part of this book, we focus primarily on two topics in designing new frameworks of multimodal pose estimation.

Toward Drift Reduction in Visual Odometry As low-cost sensors with abundant visual information, cameras are frequently seen in ground vehicles, where visual odometry (VO) has been widely used for autonomous vehicle pose estimation thanks to its constantly improving performance. However, several challenges still need to be resolved. Error accumulation or the so-called drift issue is a challenge preventing VO from being used in long-range navigation. The existing solutions for enhancing VO performance involve (i) improving VO components including feature detection, matching, outlier removal, and pose optimization; and (ii) seeking assistance from other approaches or databases [Shen et al., 2014] such as LIDAR [Zhang and Singh, 2015], global positioning systems (GPSs) [Agrawal and Konolige, 2006], digital maps [Jiang et al., 2017, Alonso et al., 2012], and inertial navigation systems (INS) [Bloesch et al., 2015, Mourikis and Roumeliotis, 2007, Lobo and Dias, 2003, Wang et al., 2014, Falquez et al., 2016, Leutenegger et al., 2015, Lupton and Sukkarieh, 2012, Forster et al., 2017, Piniés et al., 2007, Li and Mourikis, 2013, Santoso et al., 2017]. Benefiting from the self-contained property, many visual-inertial odometry (VIO) schemes have been proposed to reduce drift in VO. Loosely coupled methods [Mourikis and Roumeliotis, 2007, Falquez et al., 2016] fuse data at a higher level, where data from the inertial measurement unit (IMU) and VO are fused after being obtained; tightly coupled methods, which consider not only poses but features as state variables in estimation, generally achieve greater precision but also suffer from higher computational costs. There are two main streams in tightly coupled VIO: on the one hand, a filter-based method is proposed to estimate egomotion, camera extrinsic parameters, and the additive IMU biases in Bloesch et al. (2015). On the other hand, with optimization-based methods, pose estimation can be formulated as a non-linear least-square optimization problem that aims to minimize a cost function containing inertial error terms and reprojection error simultaneously. Leutenegger et al. (2015) have proposed an integration framework where the concepts of keyframes and marginalization are introduced to ensure real-time operation. In Forster et al. (2017), a preintegration scheme of inertial measurement between keyframes has been proposed, where a fused measurement model and error propagation expression have been derived such that the optimization could be achieved directly on-manifold. Yang and Shen (2017) have addressed the initialization and calibration problems on the fly for monocular VIO. Unfortunately, the existing methods still suffer from drift issues, which motivates us to eliminate rotation drift by introducing an absolute heading in pose estimation.

As for the deployment of orientations as supplementary information, gravity has been used as a vertical reference for vision and inertial sensor cooperation in structure from motion (SfM) methods, where the image horizon line can be determined [Lobo and Dias, 2003]. Saurer et al. (2017) have further proposed an egomotion estimation approach, where fewer point correspondences for relative motion estimation are needed by utilizing the gravity direction. To the best of the authors' knowledge, Chapter 2 of this book serves as the first attempt to discuss the heading reference-assisted pose estimation problem for ground vehicles.

Map-Aided Visual Dead-Reckoning Stereo visual odometry (SVO) and monocular visual odometry (MVO) are two popular forms of visual dead-reckoning in practice. The motion estimated from SVO drifts in six degrees of freedom, while the motion estimated from MVO drifts in seven degrees of freedom, with an additional scale drift. To correct the drift, loop closure detection, followed by a global bundle adjustment or graph optimization [Strasdat et al., 2010] step, is widely used, and impressive results have been achieved [Mur-Artal et al., 2015a]. Unfortunately, loops do not necessarily exist in practical driving conditions. But when loops do exist, the corrected motion is still a largely delayed result for the route before loop closure. Thus, the loop-closing method is not appropriate for applications where instantaneous decisions are desired, such as autonomous vehicles.

For MVO, another challenge is how to obtain the metric scale of the motion estimated from a monocular system. The most straightforward approach is to fuse information from the IMU, GPS, or other sensors [Nützi et al., 2011], which of course will increase the cost and complexity of the system. Another popular method to estimate the metric scale assumes that the camera is moving at a known fixed height over the ground [Song and Chandraker, 2014]. However, the result of this kind of method relies heavily on the accuracy of ground plane detection. Some other researchers propose to use objects with known sizes to give the absolute scale of monocular results [Davison, 2003]. Nevertheless, it is difficult to ensure that the objects appear and are detected in all frames.

To solve the drift and scale ambiguity challenge, we turn to freely available maps (such as OpenStreetMap [OSM] and Google Maps), which have plenty of information that can be used for localization. On the one hand, since street segments and the connectivity of roads can be intuitively expressed as nodes and edges of a graph model, a map can be represented by a directed graph, which is much simpler to work within a localization problem. In Brubaker et al. (2013), a graph-based representation of the map was defined, and a probabilistic map localization approach was proposed. They achieved an accuracy of 3 meters within a city-level road map by using VO measurements and OSM data as the only inputs. To reduce the high computational cost, a simplified approach that used wheel speed odometry instead of VO was proposed in Merriaux et al. (2015), and real-time performance was achieved. On the other hand, the geometric shape of road networks can be considered as a constraint to assist with position estimation. Senlet and Elgammal (2012) succeeded in localizing a mobile robot on sidewalks by using a combination of SVO and satellite map matching. Top-view images generated from stereo frames were matched with satellite maps to correct incremental drift. The shape-matching method was utilized to evaluate the alignment of different trajectories to the map in Floros et al. (2013), and a shape-matching process was considered as the measurement model of the Monte Carlo localization framework. Similarly, shape matching is also used in Chapter 3. However, there is a difference: we focus on monocular camera localization, while they worked on the stereo case. Thus, not only motion drift but also scale ambiguity are considered in Chapter 3.

1.3 Secure Estimation

1.3.1 Secure State Estimation under Cyber Attacks

Cyber attacks reduce the reliability of the system primarily by destroying the availability and integrity of sensor or actuator data. A deception attack is a typical network attack that is more difficult to be detected because an adversary can keep the attack hidden from an anomaly detector. A false data injection attack is a specific type of deception attack when the system model is known to the adversary. It is mentioned that an alternative way to achieve secure estimation is to treat the false data injection attack signal as unknown input. Although classical methods on unknown input can be used to solve some secure estimation problems, when the system suffers from an attack and unknown input simultaneously, there seem to be some limitations. The main reason is that the existing literature considers the two unknown inputs in the system model (state equation and measurement equation) are the same. Once two identical unknown inputs are set as different signals, the analysis of filter gain and stability needs to be reconsidered. Therefore, based on existing results about unknown input, how to solve the secure estimation problem for systems in the presence of random attacks and unknown input remains a challenge and serves as the main motivation of our book.

Networked-embedded sensors are ubiquitous in monitoring dynamical systems due to their low cost and ease of installation. However, they are also vulnerable to attackers owing to their limited capacity and sparse spatial deployment. Attackers might gain access to sensors and arbitrarily manipulate sensor measurements or break the communication links between sensors and system operators to inject faked information. Therefore, the secure state estimation problem of linear dynamical systems under sparse sensor attacks has been extensively studied in the past few years. In the problem setting, it is usually assumed that a group of sensors are deployed to monitor the system's dynamics, of which a subset of sensors might be compromised and their measurements arbitrarily tampered with. The problem of interest is to determine the conditions under which the system states can be reliably estimated and to design secure estimators to generate reliable estimates.

1.3.2 Secure Pose Estimation for Autonomous Vehicles

Autonomous vehicles (AVs) have recently attracted significant attention from both the academic and industrial domains. Today, many automobile manufacturers and IT companies have implemented active programs to develop AV technologies. Before AV technologies become mature and ultimately available for massive usage, their security problems should be well addressed. In 2015, hackers took control of a Jeep Cherokee and crashed it into a ditch by remotely breaking into its dashboard computer from 10 miles away. Therefore, security is always one of the most critical issues in all kinds of vehicle applications for both AVs and traditional vehicles.

AVs with a small form factor, quick responses, and the ability to operate remotely in a difficult and challenging environment have wide applications. To accomplish an autonomous task, the vehicle should have the capability to determine its pose using sensors mounted on the vehicle. Accurate pose estimation of the AV still remains a significant challenge, especially in GPS-denied environments. Small vehicles like micro aerial vehicles (MAVs) have major constraints, including limited payload carrying capacity. Therefore, a wise selection of sensor(s) and processing unit is important. Equally important is an online, real-time estimation of the pose of the vehicle. For AVs, this book aims to develop a resilient and efficient run-time estimation algorithm that provides a performance guarantee in the presence of malicious sensor attacks.

1.4 Contributions and Organization

Chapters 2 to **5** focus on multi-sensor pose estimation for ground vehicles, while **Chapters 6** to **9** formulate secure dynamic state estimation for mobile robots. Finally, **Chapter 10** concludes the book and provides several perspectives for future work. The primary contributions in the book are summarized and highlighted as follows:

- In order to suppress VO rotation drift in pose estimation, the absolute heading is used to assist VO in **Chapter 2** [Wang et al., 2018a]. By utilizing the coupling characteristics between rotation and translation in VO, the heading measurement can be cost-effectively used to benefit both rotation and translation estimation in ground vehicles. With the absolute heading, a sensor fusion framework is introduced to incorporate heading measurements into VO. In the framework, a heading reference is abstracted as a vertex in the graph model, based on which the problem is formulated as a graph optimization such that off-the-shelf back-end libraries can be utilized effortlessly. To demonstrate the effectiveness of the heading reference-assisted approach, extensive experiments have been conducted based on the KITTI dataset and self-collected data. The results are compared and discussed between pure VO and the proposed approach.
- In addition to the drift issue, the scale ambiguity of MVO is considered as a measurement uncertainty and incorporated into a unified probability distribution. A uniform Gaussian distribution (UGD) is introduced in **Chapter 3** [Jiang et al., 2017, Yang et al., 2017b] to describe measurement uncertainties in VO. The UGD model is used to generate particles representing either SVO or MVO measurements that contain drift and scale ambiguity. A parameter estimation scheme is then presented to refine the probability distribution according to sample particles. The parameter estimation result is used to generate particles iteratively, similar to the Monte Carlo localization framework. Combined with particle *saliency* representing VO measurement certainty degree and particle *consistency* denoting accordance with the map, a map-assisted localization framework is introduced to reduce the drift and scale ambiguity of VO.
- In **Chapter 4** [Jiang et al., 2018], we present a fusion approach to localize urban vehicles by integrating VO, a low-cost GPS, and a two-dimensional digital road map. Distinguished from conventional sensor fusion methods, two types of

potential functions (i.e. potential wells and potential trenches) are introduced to represent measurements and constraints, respectively. By choosing different potential functions according to data properties, data from various sensors can be integrated with intuitive understanding, while no extra map matching is required. The minimum of the fused potential, which is regarded as a position estimation, is confined such that fast minimum searching can be achieved. Experiments under realistic conditions have been conducted to validate satisfactory positioning accuracy and robustness compared to pure VO and map-matching methods.

- In **Chapter 5** [Jiang et al., 2019c], to fuse information from multiple sensors and constraints in pose estimation, non-Euclidean state space and measurement space are presented as background spaces in which states and measurements are points while constraints are subsets. With the distance defined on connected Riemannian manifolds, dynamic potential fields (DPFs) are designed accordingly to represent states, measurements, and constraints such that system noise, measurement noise, and constraint softness are modeled. Based on the DPF formulation, an information fusion scheme called Geometric Pose Estimation (Geo-PE) is presented, in which the state equation and output equation are considered mappings between state-sourced and measurement-sourced DPFs with clear probabilistic implications; the state-sourced and measurement-sourced DPFs are then projected to constraints along the derivative of constraint-sourced DPF. Constraints are thus inherently considered compared to conventional sensor fusion approaches. The approximated version of Geo-PE is then presented, catering to various systems without explicit analytic expression. Geo-PE has been developed for ground vehicles equipped with SVO, an attitude and heading reference system, and OSM so that the pose transformation from dead-reckoning, heading in the East-North-Up frame, and road map in the East-North plane are incorporated for estimating the vehicle pose without rotation drift. Experiments have been conducted based on public KITTI sequences and self-collected Nanyang Technological University (NTU) dataset to validate the performance of the proposed approach.
- In **Chapter 6** [Liu et al., 2019b], we present a filter-based secure dynamic pose estimation approach such that the vehicle pose can be resilient under possible sensor attacks. Our estimator coincides with the conventional Kalman filter when all sensors on AVs are benign. If less than half of the measurement states are compromised by randomly occurring deception attacks, it still gives stable estimates of the pose states: i.e. an upper bound is guaranteed for the estimation error covariance. The pose estimation results with single and multiple attacks on the testing route validate the effectiveness and robustness of the proposed approach.

- In **Chapter 7** [Liu et al., 2021], we introduce a pose estimation approach for ground vehicles under randomly occurring deception attacks. By modeling attacks as signals added to measurements with a certain probability, the attack model is presented and incorporated into the existing process and measurement equations of ground vehicle pose estimation based on multi-sensor fusion. An unscented Kalman filter (UKF)-based secure pose estimator is then established to generate a stable estimate of the vehicle pose states; i.e. an upper bound is guaranteed for the estimation error covariance. The simulation and experiments are conducted on a simple but effective single-input-single-output dynamic system and the ground vehicle model to show the effectiveness of UKF-based secure pose estimation.
- In **Chapter 8** [Liu et al., 2017], we consider the problem of estimating the state of a linear time-invariant Gaussian system in the presence of sparse integrity attacks. The attacker can control p out of m sensors and arbitrarily change the measurements. Under mild assumptions, we can decompose the optimal Kalman estimate as a weighted sum of local state estimates, each of which is derived using only the measurements from a single sensor. Furthermore, we introduce a convex optimization-based approach, instead of the weighted sum approach, to combine the local estimate into a more secure state estimate. It is shown that our proposed estimator coincides with the Kalman estimator with a certain probability when all sensors are benign, and we provide a sufficient condition under which the estimator is stable against the (p, m)-sparse attack when p sensors are compromised. A numerical example is provided to illustrate the performance of the state estimation scheme.
- In **Chapter 9** [Jiang et al., 2019a], we focus on the problem of secure attitude estimation for AVs. Based on the established attitude and heading reference system (AHRS) measuring model and the attack model, we have decomposed the optimal Kalman estimate into a linear combination of local state estimates. We then introduce a convex optimization-based approach, instead of the weighted sum approach, to combine the local estimate into a more secure estimate. It is shown that the secure estimator coincides with the Kalman estimator with a certain probability when there is no attack and can be stable when p elements of the model state are compromised. Simulations have been conducted to validate the secure filter under single and multiple measurement attacks.

Part I

Multimodal Perception in Vehicle Pose Estimation

2

Heading Reference-Assisted Pose Estimation

Visual odometry (VO), a promising pose estimation approach, has been extensively used in robots and autonomous ground vehicles [Cheng et al., 2005, Nistér et al., 2006]. Although prominent achievements have been made in test datasets [Geiger et al., 2012], it is still challenging to implement VO in environments with poor illumination conditions, insufficient features, and dynamic scenes.

Translation error and rotation error, both of which cause error accumulation in VO, are closely coupled. A small drift in rotation may result in abysmal VO performance. This chapter works on assisting VO by improving rotation estimation with an orientation reference. Although visual-inertial odometry (VIO) has performed satisfactorily, components in VIO are based on dead-reckoning, which is inherently sensitive to drifting issues. Thus, it is necessary to define an orientation in the global frame to provide an absolute reference for drift suppression. The recent development of micro-electromechanical technologies enables compact, precise, cost-efficient attitude and heading reference systems (AHRSs) that provide accurate and timely orientation estimations. In this chapter, a pose estimation framework is proposed that is differentiated from VIO: an absolute heading reference is used to assist VO such that robust, accurate vehicle localization can be attained.

The remainder of the chapter is organized as follows. The principles of stereo visual odometry (SVO), heading reference sensors, and the graph optimization problem are introduced in Section 2.1. Section 2.2 provides two levels of abstraction models involving heading references. In Section 2.3, a general framework of pose estimation with heading references is proposed. Sections 2.4 and 2.5 detail the implementation of the proposed approach and demonstrate the experimental results. Finally, 2.6 concludes the chapter.

Multimodal Perception and Secure State Estimation for Robotic Mobility Platforms, First Edition.
Xinghua Liu, Rui Jiang, Badong Chen, and Shuzhi Sam Ge.

2.1 Preliminaries

2.1.1 Stereo Visual Odometry

As an integrated sensor, VO can be modeled as [Scaramuzza and Fraundorfer, 2011]

$$\mathrm{C}_{k+1} = \mathrm{C}_k \mathrm{T}_{k,k+1} \tag{2.1}$$

where $\mathrm{C}_k = [\mathrm{R}_k|\mathbf{t}_k] \in \mathrm{SE}(3)$ denotes the camera pose matrix at time instant k; $\mathrm{T}_{k,k+1} \in \mathrm{SE}(3)$, which is the measurement, denotes the transformation matrix between poses from k to $k+1$; and $\mathrm{R}_k \in \mathrm{SO}(3)$ and $\mathbf{t}_k \in \mathbb{R}^3$ are the rotation matrix and translation vector at k, respectively.

Taking VO as an integrated module may be insufficient in scenarios where feature-level optimization is necessary. With an image pair from a stereo camera system, the coordinates of 3D feature points (or *landmarks*) can be recovered. The 3D-3D correspondences can be utilized to calculate camera motion. Suppose the landmark is represented as $\mathbf{x}_\mathrm{L} = [x_\mathrm{L}, y_\mathrm{L}, z_\mathrm{L}]^\top$ in the global frame. The ideal measurement equation that projects landmarks to homogeneous image coordinates on the focal plane are given as follows [Geiger et al., 2011a] and illustrated in Figure 2.1

$$\begin{bmatrix} u \\ v \\ 1 \end{bmatrix} = \begin{bmatrix} f & 0 & c_u \\ 0 & f & c_v \\ 0 & 0 & 1 \end{bmatrix} \left(\begin{bmatrix} {}^\mathrm{G}_\mathrm{C}\mathrm{R}_k & {}^\mathrm{G}_\mathrm{C}\mathbf{t}_k \end{bmatrix} \begin{bmatrix} x_\mathrm{L} \\ y_\mathrm{L} \\ z_\mathrm{L} \\ 1 \end{bmatrix} - \begin{bmatrix} s \\ 0 \\ 0 \end{bmatrix} \right) \tag{2.2}$$

with focal length f, projection center $[c_u, c_v]^\top$, rotation matrix and translation vector from the global frame to the camera frame ${}^\mathrm{G}_\mathrm{C}\mathrm{R}_k$, and ${}^\mathrm{G}_\mathrm{C}\mathbf{t}_k$. The shift $s = 0$ equals the baseline of a stereo camera.

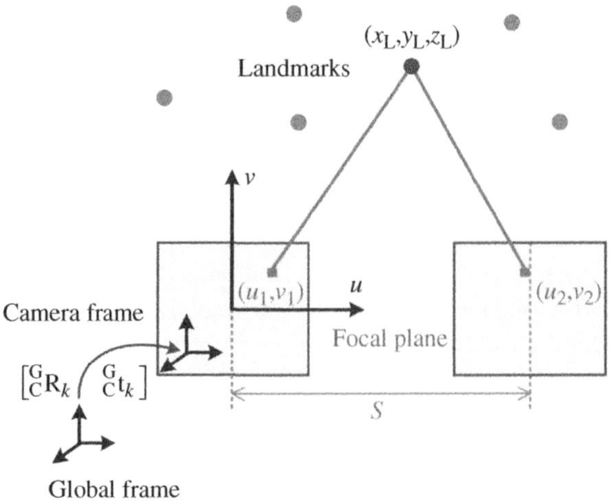

Figure 2.1 Stereo 3D-2D projection from Euclidean space to the focal plane.

2.1.2 Heading Reference Sensors

Herein, the rotation from the camera frame to the East-North-Up (ENU) frame is taken as the heading measurement for ground vehicles. The rotation between two frames can be estimated by measuring the ambient vector field in different frames. Without loss of generality, let us suppose that the origins of all frames coincide with each other. The static ambient vector field $\mathbf{v} \in \mathbb{R}^3$ is written as ${}_{\mathrm{G}}\mathbf{v}$ in the global frame. By measuring $\mathbf{v}$ in the body frame as ${}_{\mathrm{B}}\mathbf{v}$, the heading from the body frame to the global frame can be represented as a rotation matrix ${}^{\mathrm{B}}_{\mathrm{G}}\mathrm{R}$ that satisfies ${}_{\mathrm{G}}\mathbf{v} = {}^{\mathrm{B}}_{\mathrm{G}}\mathrm{R}\,{}_{\mathrm{B}}\mathbf{v}$. The angle and unified axis of the rotation are derived as

$$\theta = \arccos\left(\frac{{}_{\mathrm{B}}\mathbf{v} \cdot {}_{\mathrm{G}}\mathbf{v}}{\|{}_{\mathrm{B}}\mathbf{v}\|\|{}_{\mathrm{G}}\mathbf{v}\|}\right) \tag{2.3}$$

$$\mathbf{a} = \frac{{}_{\mathrm{B}}\mathbf{v} \times {}_{\mathrm{G}}\mathbf{v}}{\|{}_{\mathrm{B}}\mathbf{v} \times {}_{\mathrm{G}}\mathbf{v}\|.} \tag{2.4}$$

The rotation matrix can be obtained using an exponential map as

$${}^{\mathrm{B}}_{\mathrm{G}}\mathrm{R} = \exp\left([\mathbf{a}]_\times \theta\right) = \mathbf{I}_3 + \sin\theta[\mathbf{a}]_\times + (1 - \cos\theta)[\mathbf{a}]_\times^2 \tag{2.5}$$

where $\mathbf{I}_3$ is the identity matrix of size 3; $[\mathbf{a}]_\times$ denotes the skew-symmetric matrix (or *cross-product matrix*) for vector $\mathbf{a}$. Unfortunately, by noticing the norm constraint $\|{}_{\mathrm{B}}\mathbf{v} = {}_{\mathrm{G}}\mathbf{v}\|$, equation system ${}_{\mathrm{G}}\mathbf{v} = {}^{\mathrm{B}}_{\mathrm{G}}\mathrm{R}\,{}_{\mathrm{B}}\mathbf{v}$ is underdetermined. In other words, at least two vector measurements are required for the same time to determine a unique ${}^{\mathrm{B}}_{\mathrm{G}}\mathrm{R}$. A general method to obtain the solution would be minimizing the loss function $J({}^{\mathrm{B}}_{\mathrm{G}}\mathrm{R}) = \frac{1}{2}\sum w_k \|{}_{\mathrm{G}}\mathbf{v} - {}^{\mathrm{B}}_{\mathrm{G}}\mathrm{R}\,{}_{\mathrm{B}}\mathbf{v}\|^2$ from multiple sensors, where w_k denotes the weight for sensor k (known as Wahba's problem) [Wahba, 1965].

Since the motion range of most applications is limited in a small-scale space where Earth's magnetic field and gravitational field can be considered as constant, a magnetometer and accelerometer are typical components in heading reference sensors. In practice, a low-pass filter is required to extract the gravity from accelerometer values such that the influence of motion can be filtered out. To achieve high-frequency data processing at up to 150 Hz, QUEST (QUaternion ESTimator) [Shuster and Oh, 1981] has been extensively used, which gives direct quaternion sub-optimal estimation as

$$\mathbf{q} = \frac{1}{\sqrt{1 + \mathbf{p}^\top \mathbf{p}}} \begin{bmatrix} \mathbf{p} \\ 1 \end{bmatrix} \tag{2.6}$$

where $\mathbf{p} = [(\sum w_k + \sigma)\mathbf{I} - \mathbf{S}]^{-1}\mathbf{Z}$, $\sigma = \mathrm{tr}(\mathbf{B})$, $\mathbf{S} = \mathbf{B} + \mathbf{B}^\top$, $\mathbf{B} = \sum w_k \left({}_{\mathrm{G}}\mathbf{v}_{\mathrm{B}}\mathbf{v}^\top\right)$, and $\mathbf{Z} = [B_{23} - B_{32}, B_{31} - B_{13}, B_{12} - B_{21}]^\top$.

2.1.3 Graph Optimization on a Manifold

A graph is a frequently used model to represent measurements and states. Let $\mathbf{x}_i$ describe the state of vertex i, $\mathbf{h}_{ij}(\mathbf{x}_i, \mathbf{x}_j)$ be the ideal measurement equation, and $\mathbf{z}_{ij}$ denote the actual measurement between vertex i and vertex j. With the existence of measurement error, we can always define an error function as $\mathbf{e}_{ij}(\mathbf{z}_{ij}, \mathbf{x}_i, \mathbf{x}_j) = \mathbf{z}_{ij} \ominus \mathbf{h}_{ij}(\mathbf{x}_i, \mathbf{x}_j)$, where $\ominus$ is an operator measuring the difference between the ideal and actual measurements. To simplify the notation, we let $\mathbf{x} = [\mathbf{x}_1^\top, \cdots, \mathbf{x}_n^\top]^\top$ be the parameter vector indicating n vertices and $\mathbf{x}_{ij} = [\mathbf{x}_i^\top, \mathbf{x}_j^\top]^\top$. The graph optimization problem aims to obtain the optimal $\mathbf{x}$ as

$$\mathbf{x}^* = \arg\min_{\mathbf{x}} \sum_{<i,j>\in C} \underbrace{\mathbf{e}_{ij}(\mathbf{z}_{ij}, \mathbf{x}_{ij})^\top \boldsymbol{\Omega}_{ij} \mathbf{e}_{ij}(\mathbf{z}_{ij}, \mathbf{x}_{ij})}_{\mathbf{F}_{ij}} \tag{2.7}$$

where C denotes the index set containing measurements and $\boldsymbol{\Omega}_{ij}$ represents the information matrix of measurement $\mathbf{z}_{ij}$.

As it is difficult to find an analytical solution to Eq. (2.7) due to non-linearity, iterations are typically required until a numerical solution is obtained. In Euclidean space, given an initial guess $\mathring{\mathbf{x}}_{ij}$, the error function around $\mathring{\mathbf{x}}_{ij}$ can be approximated as

$$\mathbf{e}_{ij}(\mathring{\mathbf{x}}_{ij} + \Delta\mathbf{x}_{ij}) \simeq \mathbf{e}_{ij}(\mathring{\mathbf{x}}_{ij}) + \mathbf{J}_{ij}\Delta\mathbf{x}_{ij} \tag{2.8}$$

where $\mathbf{J}_{ij}$ is the Jacobian of $\mathbf{e}_{ij}$ at $\mathring{\mathbf{x}}_{ij}$. The term $\mathbf{z}_{ij}$ in the cost function is omitted as we are optimizing $\mathbf{x}_{ij}$ instead of the measurement. Taking Eq. (2.8) into $\mathbf{F}_{ij}$ leads to

$$\mathbf{F}_{ij}(\mathring{\mathbf{x}}_{ij} + \Delta\mathbf{x}_{ij}) = \underbrace{\mathbf{e}_{ij}^\top(\mathring{\mathbf{x}}_{ij})\boldsymbol{\Omega}_{ij}\mathbf{e}_{ij}(\mathring{\mathbf{x}}_{ij})}_{c_{ij}} +$$

$$\underbrace{2\mathbf{e}_{ij}^\top(\mathring{\mathbf{x}}_{ij})\boldsymbol{\Omega}_{ij}\mathbf{J}_{ij}}_{\mathbf{b}_{ij}^\top}\Delta\mathbf{x}_{ij} + \Delta\mathbf{x}_{ij}^\top\underbrace{\mathbf{J}_{ij}^\top\boldsymbol{\Omega}_{ij}\mathbf{J}_{ij}}_{\mathbf{H}_{ij}}\Delta\mathbf{x}_{ij} \tag{2.9}$$

$$= c_{ij} + 2\mathbf{b}_{ij}^\top\Delta\mathbf{x}_{ij} + \Delta\mathbf{x}_{ij}^\top\mathbf{H}_{ij}\Delta\mathbf{x}_{ij}. \tag{2.10}$$

The summed cost function can be written as

$$\mathbf{F}(\mathring{\mathbf{x}} + \Delta\mathbf{x}) = \sum_{<i,j>\in C} \mathbf{F}_{ij}(\mathring{\mathbf{x}}_{ij} + \Delta\mathbf{x}_{ij}) \tag{2.11}$$

$$\simeq c + 2\mathbf{b}^\top\Delta\mathbf{x} + \Delta\mathbf{x}^\top\mathbf{H}\Delta\mathbf{x} \tag{2.12}$$

where $c = \sum c_{ij}$, $\mathbf{b} = \sum \mathbf{b}_{ij}$, $\mathbf{H} = \sum \mathbf{H}_{ij}$. Taking the derivative of $\mathbf{F}(\mathring{\mathbf{x}} + \Delta\mathbf{x})$ with respect to $\Delta\mathbf{x}$ leads to a sparse linear system $\mathbf{H}\Delta\mathbf{x}^* = -\mathbf{b}$, from which $\Delta\mathbf{x}^*$ can be obtained as the optimal increment to the initial guess of the whole state $[\Delta\mathbf{x}_1^{*\top}, \cdots, \Delta\mathbf{x}_M^{*\top}]^\top$ with M vertices. Then the linearized optimal solution is obtained as

$$\mathbf{x}^* = \mathring{\mathbf{x}} + \Delta\mathbf{x}^*. \tag{2.13}$$

When states are parameterized in non-Euclidean space, a similar optimization procedure can be carried out on a manifold, which is regarded as Euclidean space locally but not globally. To represent optimization problem on a manifold, an operator $\boxplus$ needs to be defined to replace the simple addition such that the state after a small perturbation $\mathbf{x} \boxplus \Delta\mathbf{x}$ is still on the manifold. The error function on the manifold can be modified as

$$\mathbf{e}_{ij}(\mathring{\mathbf{x}}_{ij} \boxplus \Delta\mathbf{x}_{ij}) \simeq \mathbf{e}_{ij}(\mathring{\mathbf{x}}_{ij}) + \tilde{\mathbf{J}}_{ij}\Delta\mathbf{x}_{ij} \tag{2.14}$$

where $\tilde{\mathbf{J}}_{ij} = \frac{\partial \mathbf{e}_{ij}(\mathring{\mathbf{x}}_{ij} \boxplus \Delta\mathbf{x}_{ij})}{\partial \Delta\mathbf{x}_{ij}}|_{\Delta\mathbf{x}_{ij}=0}$. In a VO system, to maintain rotational constraints, the small rotation increment is expressed minimally using Euler angles. To avoid singularity, the states are represented in over-parameterized space SE(3). Accordingly, given an initial guess $\mathring{\mathbf{x}}$, the states can be updated from $\mathbf{x}^* = \mathring{\mathbf{x}} \boxplus \Delta\mathbf{x}^*$ after obtaining the optimal increment $\Delta\mathbf{x}^*$.

To facilitate fast implementation, several computing packages and libraries have been developed to solve the graph optimization problem [Kaess et al., 2007, Kümmerle et al., 2011, Konolige and Garage, 2010].

2.2 Abstraction Model of Measurement with a Heading Reference

The graph model must be redesigned when heading measurements are brought in. In this section, we introduce two types of graph abstraction models, both of which incorporate a heading reference and heading measurements into the visual pose estimation framework such that the existing graph models are generalized. The loosely coupled model considers VO a black box whose input and output are stereo images and a transformation matrix, respectively. At a lower abstract level, the tightly coupled model takes 2D landmarks on image planes as observations, enabling us to consider constraints between landmarks.

We consider the visual pose estimation problem for ground vehicles by graph optimization, where vehicle translation and rotation are unknown parameters to be solved. In other words, we aim to optimize the camera pose C_k at time k, given measurement set $\{\mathbf{z}_{ij}\}, \forall <i,j> \in C_k$, where C_k denotes the index set containing all available measurements until time k.

2.2.1 Loosely Coupled Model

The loosely coupled model is illustrated in Figure 2.2(a). Let $\mathbf{x} = [\mathbf{x}_0^\top, \mathbf{x}_1^\top, \cdots, \mathbf{x}_N^\top]^\top$ be the state, where $\mathbf{x}_0 \in \mathbb{R}^3$ denotes a heading reference in a three-dimensional unit quaternion vector and $\mathbf{x}_1, \cdots, \mathbf{x}_N \in \mathbb{R}^6$ are camera poses vectors with translation and rotation components.[1] Let $\mathbf{z}_{ij}$ be the measurement, where

1 We show vector dimension with a slight abuse of notation. Actually, the range of elements in unit quaternion does not cover $\mathbb{R}$. The same notation goes to $\mathbf{x}_1, \cdots, \mathbf{x}_N, \mathbf{z}_{0k}, \mathbf{z}_{k,k+1}$, and overloading functions `v2t`, `t2v`.

$\mathbf{z}_{0k} \in \mathbb{R}^3$ represents rotation from the ENU frame to the camera frame in the unit quaternion vector, and $\mathbf{z}_{k,k+1} \in \mathbb{R}^6$ denotes the transformation vector between consecutive poses at times k and $k+1$ measured by VO.

By defining an overloading function $\mathtt{v2t} : \mathbb{R}^3 \rightarrow SO(3); \mathbb{R}^6 \rightarrow SE(3)$ that converts the vector representation to a rotation or transformation matrix, we write $\mathrm{R}_0 = \mathtt{v2t}(\mathbf{x}_0)$, $\mathrm{C}_k = \mathtt{v2t}(\mathbf{x}_k)$, $\mathrm{R}_{0k} = \mathtt{v2t}(\mathbf{z}_{0k})$, and $\mathrm{T}_{k,k+1} = \mathtt{v2t}(\mathbf{z}_{k,k+1})$. The ideal measurement equations for VO and the heading are

$$\mathbf{h}^{\mathrm{VO}}(\mathrm{C}_k, \mathrm{C}_{k+1}) = \mathrm{C}_k^{-1}\mathrm{C}_{k+1} \tag{2.15}$$

$$\mathbf{h}^{\mathrm{HR}}(\mathrm{R}_0, \mathrm{C}_k) = \mathrm{R}_0^{-1}\mathrm{R}_k. \tag{2.16}$$

The error functions are defined as

$$\begin{aligned}\mathbf{e}^{\mathrm{VO}}(\mathrm{C}_k, \mathrm{C}_{k+1}) &= \mathbf{z}_{k,k+1} \ominus \mathbf{h}^{\mathrm{VO}}(\mathrm{C}_k, \mathrm{C}_{k+1})\\ &= \mathtt{t2v}\left(\{\mathbf{h}^{\mathrm{VO}}(\mathrm{C}_k, \mathrm{C}_{k+1})\}^{-1}\mathrm{T}_{k,k+1}\right)\end{aligned} \tag{2.17}$$

$$\begin{aligned}\mathbf{e}^{\mathrm{HR}}(\mathrm{R}_0, \mathrm{C}_k) &= \mathbf{z}_{0k} \ominus \mathbf{h}^{\mathrm{HR}}(\mathrm{R}_0, \mathrm{C}_k)\\ &= \mathtt{t2v}\left(\{\mathbf{h}^{\mathrm{HR}}(\mathrm{R}_0, \mathrm{C}_k)\}^{-1}\mathrm{R}_{0k}\right)\end{aligned} \tag{2.18}$$

where the function $\mathtt{t2v} : SO(3) \rightarrow \mathbb{R}^3; SE(3) \rightarrow \mathbb{R}^6$ converts the rotation matrix to a unit quaternion vector for a particular rotation while the translation vector remains unchanged. For loosely coupled VO, the measurement information matrix $\mathbf{\Omega}^{\mathrm{VO}}$ between neighboring frames is modeled as constant in this work.

2.2.2 Tightly Coupled Model

The tightly coupled model is demonstrated in Figure 2.2(b). Besides the heading measurements in the loosely coupled model, two other types of measurements have been considered: landmark projection $\mathbf{z}_{\mathrm{L}ik}$ measures 3D-2D projection for landmark i at time k, while landmark location $\mathbf{z}_{\mathrm{L}i\mathrm{L}j}$ constrains the three-dimensional distance between landmarks $\mathbf{x}_{\mathrm{L}i}$ and $\mathbf{x}_{\mathrm{L}j}$.

Landmark i in 3D is parameterized as $\mathbf{x}_{\mathrm{L}i} = [x_{\mathrm{L}i}, y_{\mathrm{L}i}, z_{\mathrm{L}i}]^{\mathsf{T}}$. According to epipolar geometry, projected coordinates v on image planes are identical for parallel cameras. Thus we extend the measurement equation (2.2) to the stereo case such that

$$\mathbf{h}^{\mathrm{LM}}(\mathbf{x}_{\mathrm{L}i}, \mathrm{C}_k) = \begin{bmatrix} u_{i1} & v_{i1} & u_{i2} \end{bmatrix}^{\mathsf{T}} \tag{2.19}$$

where

$$\begin{bmatrix} u_{i1} \\ v_{i1} \\ 1 \end{bmatrix} = \begin{bmatrix} f & 0 & c_u \\ 0 & f & c_v \\ 0 & 0 & 1 \end{bmatrix} \left(\mathrm{C}_k \begin{bmatrix} x_{\mathrm{L}i} \\ y_{\mathrm{L}i} \\ z_{\mathrm{L}i} \\ 1 \end{bmatrix} \right) \tag{2.20}$$

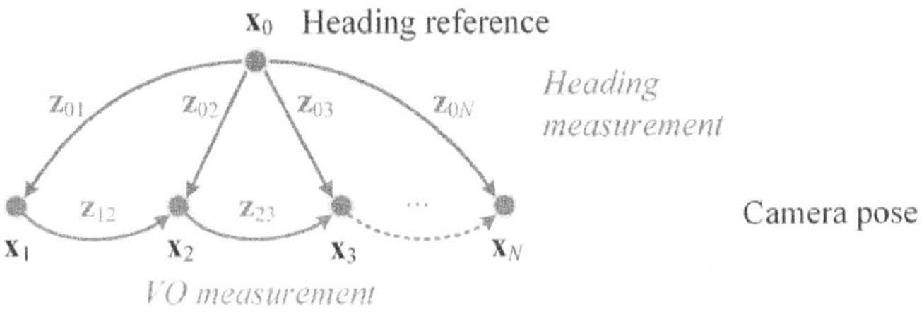

(a) Loosely coupled abstraction model

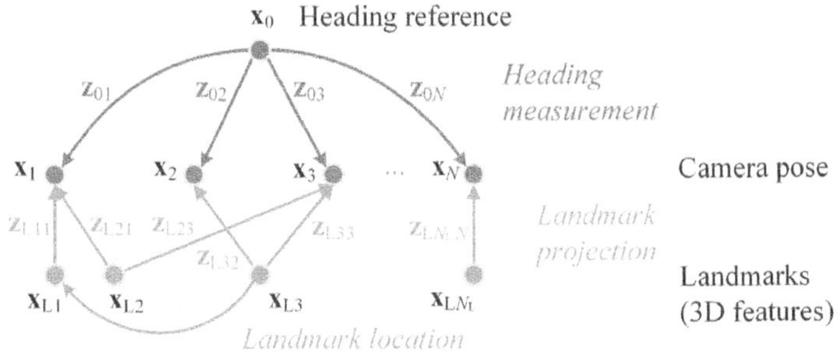

(b) Tightly coupled abstraction model

Figure 2.2 Abstraction of measurement and graph formulation.

$$\begin{bmatrix} u_{i2} \\ v_{i2} \\ 1 \end{bmatrix} = \begin{bmatrix} f & 0 & c_u \\ 0 & f & c_v \\ 0 & 0 & 1 \end{bmatrix} \left(\mathrm{C}_k \begin{bmatrix} x_{\mathrm{L}i} \\ y_{\mathrm{L}i} \\ z_{\mathrm{L}i} \\ 1 \end{bmatrix} - \begin{bmatrix} s \\ 0 \\ 0 \end{bmatrix} \right) \tag{2.21}$$

and $\mathrm{C}_k = \mathtt{v2t}(\mathbf{x}_k) = [\mathrm{R}_k|\mathbf{t}_k]$. Parameters f, c_u, c_v, s are camera-dependent and are explained in Section 2.1.1. The error function for landmark projection is defined as

$$\mathbf{e}^{\mathrm{LM}}(\mathbf{x}_{\mathrm{L}i}, \mathrm{C}_k) = \mathbf{z}_{\mathrm{L}ik} - \mathbf{h}^{\mathrm{LM}}(\mathbf{x}_{\mathrm{L}i}, \mathrm{C}_k) \tag{2.22}$$

where $\mathbf{z}_{\mathrm{L}ik} = [\tilde{u}_{i1}, \tilde{v}_{i1}, \tilde{u}_{i2}]^{\mathrm{T}}$ denotes the projected coordinates vector for the stereo pair.

Sometimes landmark location constraints are desired to represent spatial relation among landmarks. The error function for landmark location is

$$\mathbf{e}^{\mathrm{LC}}(\mathbf{x}_{\mathrm{L}i}, \mathbf{x}_{\mathrm{L}j}) = \mathbf{z}_{\mathrm{L}i\mathrm{L}j} - \mathbf{h}^{\mathrm{LC}}(\mathbf{x}_{\mathrm{L}i}, \mathbf{x}_{\mathrm{L}j}) \tag{2.23}$$

where $\mathbf{z}_{\mathrm{L}i\mathrm{L}j} \in \mathbb{R}^3$ and $\mathbf{h}^{\mathrm{LC}}(\mathbf{x}_{\mathrm{L}i}, \mathbf{x}_{\mathrm{L}j}) = \mathbf{x}_{\mathrm{L}i} - \mathbf{x}_{\mathrm{L}j}$. In the tightly coupled model, information matrices $\mathbf{\Omega}^{\mathrm{LM}}$ and $\mathbf{\Omega}^{\mathrm{LC}}$ are both set to diagonal matrices.

2.2.3 Structure of the Abstraction Model

The optimization-based state estimation problem became computationally feasible when the sparse structure of the problem was discovered. The proposed abstraction model brings new types of constraints such as heading measurements, which may cause weaker sparsity of the abstraction model. As the heading reference vertex can be considered as a subset of the camera pose vertices, the Jacobian in Eq. (2.8) has the form as

$$\mathbf{J}_{ij} = \begin{bmatrix} \cdots & \underbrace{\frac{\partial \mathbf{e}_{ij}}{\partial \mathbf{x}_i}}_{i\text{th element}} & \cdots & \underbrace{\frac{\partial \mathbf{e}_{ij}}{\partial \mathbf{x}_j}}_{j\text{th element}} & \cdots \end{bmatrix} \tag{2.24}$$

where all zero items are omitted. The structure of matrix $\mathbf{H}_{ij}$ is

$$\mathbf{H}_{ij} = \begin{bmatrix} \ddots & & & & \\ & \frac{\partial \mathbf{e}_{ij}}{\partial \mathbf{x}_i}^{\mathsf{T}} \boldsymbol{\Omega}_{ij} \frac{\partial \mathbf{e}_{ij}}{\partial \mathbf{x}_i} & \cdots & \frac{\partial \mathbf{e}_{ij}}{\partial \mathbf{x}_i}^{\mathsf{T}} \boldsymbol{\Omega}_{ij} \frac{\partial \mathbf{e}_{ij}}{\partial \mathbf{x}_j} & \\ & \vdots & & \vdots & \\ & \frac{\partial \mathbf{e}_{ij}}{\partial \mathbf{x}_j}^{\mathsf{T}} \boldsymbol{\Omega}_{ij} \frac{\partial \mathbf{e}_{ij}}{\partial \mathbf{x}_i} & \cdots & \frac{\partial \mathbf{e}_{ij}}{\partial \mathbf{x}_j}^{\mathsf{T}} \boldsymbol{\Omega}_{ij} \frac{\partial \mathbf{e}_{ij}}{\partial \mathbf{x}_j} & \\ & & & & \ddots \end{bmatrix} \tag{2.25}$$

and the matrix $\mathbf{H}$ has the structure shown in Figure 2.3, from which we know that the proposed graph model remains sparse. The newly introduced heading measurements and VO measurements lead to block tridiagonal matrices in the loosely coupled model. In the tightly coupled model, landmark location constraints may slightly weaken the sparsity of generated graph, but we are not able to predict the sparsity unless measurements have been taken. The selection of an abstract level is problem-dependent by considering sensors, measurements, constraints, and computing power. A comparative analysis between these two models can be found in Sections 2.4 and 2.5.

2.2.4 Vertex Removal in the Abstraction Model

In the proposed model, hundreds of landmark vertices are generated for each frame in the tightly coupled model. Even in the loosely coupled model, the number of vertices increases at a linear rate. It is impracticable to retain all vertices and edges during the real-time optimization process. As the optimization problem is non-linear and non-convex, only a qualitative discussion with theoretical assumptions will be presented here to help better explain the simulative and experimental results.

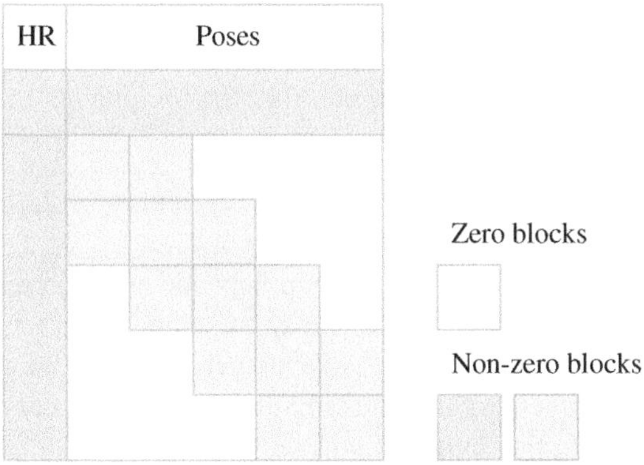

(a) Structure of matrix **H** in loosely coupled abstraction model

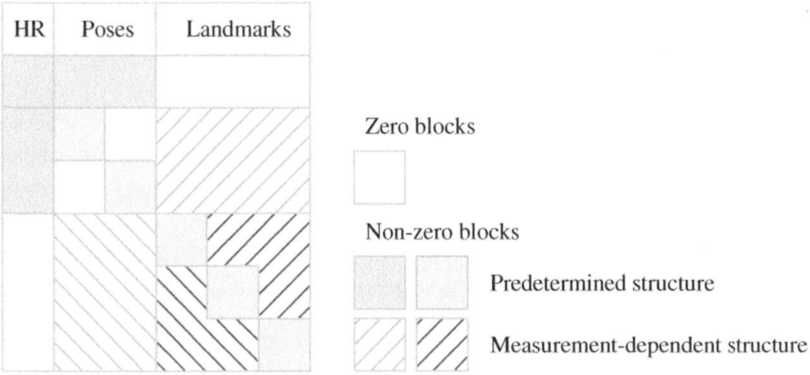

(b) Structure of matrix **H** in tightly coupled abstraction model

Figure 2.3 Illustration of **H** matrices for the loosely coupled and tightly coupled models, where *heading reference* is abbreviated *HR*. (a) A block tridiagonal matrix is obtained in the lower-right part due to constraints between consecutive VO poses. (b) Non-zero blocks exist mainly due to landmark projection measurements. In addition, landmark location constraints may cause non-zero blocks in the lower-right part, but the sparsity will remain as the number of landmark location constraints is small compared to the total number of landmarks.

Given a graph model with a local minimum $\mathbf{x}^* = \arg\min_{\mathbf{x}} \sum_{<i,j>\in C} \mathbf{e}_{ij}{}^{\mathsf{T}} \boldsymbol{\Omega}_{ij} \mathbf{e}_{ij}$, we consider an augmented optimization problem

$$\mathbf{x}^*_{\text{aug}} = \arg\min_{\mathbf{x}_{\text{aug}}} \left(\sum_{<i,j>\in C} \mathbf{e}_{ij}{}^{\mathsf{T}} \boldsymbol{\Omega}_{ij} \mathbf{e}_{ij} + \sum_{<i,j>\in C_{\text{new}}} \mathbf{e}_{ij}{}^{\mathsf{T}} \boldsymbol{\Omega}_{ij} \mathbf{e}_{ij} \right) \tag{2.26}$$

where $\mathbf{x}_{\text{aug}} = [\mathbf{x}^{\mathsf{T}}, \mathbf{x}_{\text{new}}^{\mathsf{T}}]^{\mathsf{T}}$, and C_{new} only contains indices connecting to $\mathbf{x}_{\text{new}}$.

Let us expand the error terms as $\mathbf{e}_{ij} = \mathbf{e}_{ij}(\mathbf{z}_{ij}, \mathbf{x}_i, \mathbf{x}_j) = \mathbf{e}_{ij}(\mathbf{x}_i, \mathbf{x}_j)$. The cost function in the augmented problem can be separated into two parts, where each is a quadratic form with respect to $\mathbf{e}_{ij}(\mathbf{x}_i, \mathbf{x}_j)$. It is obvious that for the augmented problem, $\mathbf{x}^*_{\text{aug}} = [\mathbf{x}^{*\top}, \mathbf{x}^{*\top}_{\text{new}}]^\top$ is a local minimum where

$$\mathbf{x}^*_{\text{new}} = \arg\min_{\mathbf{x}_{\text{new}}} \sum_{<i,j>\in C_{\text{new}}} \mathbf{e}_{ij}{}^\top \boldsymbol{\Omega}_{ij} \mathbf{e}_{ij}. \tag{2.27}$$

Whether the local minimum is also the global minimum depends on the error function $\mathbf{e}_{ij}$. In a trivial case where $\mathbf{e}_{ij}$ is linear with $\mathbf{x}_i$ and $\mathbf{x}_j$, the augmented problem is convex, which makes optimization effortless. Unfortunately, as discussed in Sections 2.2.1 and 2.2.2, most error functions in pose estimation are non-linear and non-convex. Thus the initial value in the optimization should be closely selected such that the numerical method is less like to fall into the local optimum.

2.3 Heading Reference-Assisted Pose Estimation (HRPE)

As illustrated in Figure 2.4, this section presents the pose estimation scheme, where landmarks on stereo images and headings are regarded as raw measurements from the cameras and heading reference sensors, respectively. For a loosely coupled model, egomotion estimation based on image feature correspondence (SVO) must first be performed to obtain VO measurements. For a tightly coupled model, the matched feature points are directly abstracted as vertices in the pose graph. Each time measurements are acquired, the updated graph is optimized with respect to the cost function; then, vertices and edges out of the sliding window are discarded during graph maintenance. For convenient representations, we label the pose estimation approaches based on the loosely coupled and tightly coupled model *LC-HRPE* and *TC-HRPE*, respectively, in subsequent sections.

2.3.1 Initialization

Initialization aims to acquire an accurate estimation of the heading reference. By taking advantage of insignificant VO drift, the heading references and camera poses in the first few frames are optimized simultaneously to minimize the influence of heading measurement random errors. During the rest of the optimization process, the heading reference vertex is fixed.

2.3.2 Graph Optimization

As heading measurements are received much more often than input is received from the cameras, filtering and down-sampling can be done in preprocessing for

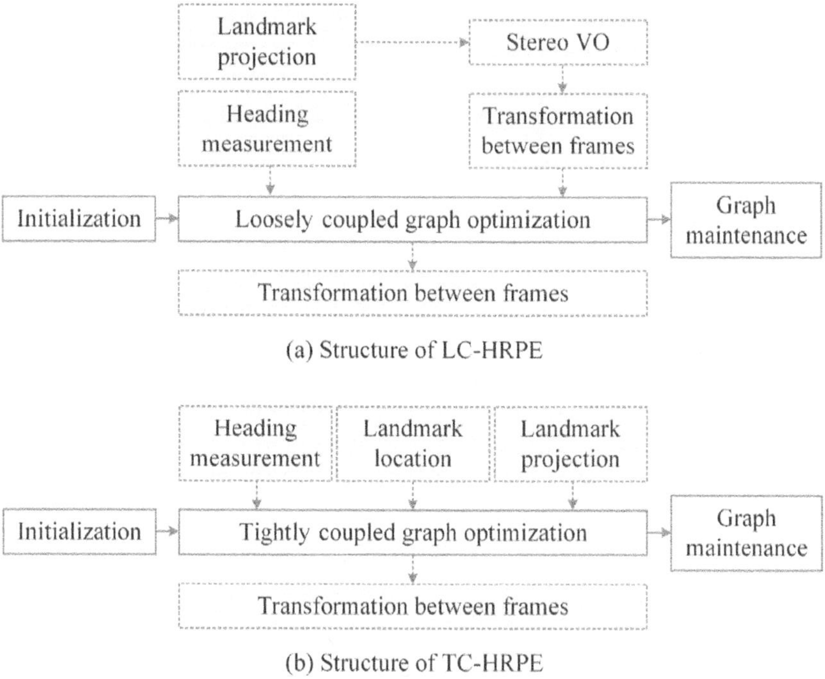

Figure 2.4 HRPE scheme. Solid arrows represent evolution over time for a single estimation, while dotted arrows denote information flow.

noise reduction and data synchronization. It is assumed that all measurements are synchronized before graph optimization. The optimization goals in LC-HRPE and TC-HRPE are written as

$$\mathbf{x}^* = \arg\min_{\mathbf{x}} \sum_{<i,j>\in \mathcal{C}^{\mathrm{VO}}} \mathbf{e}^{\mathrm{VO}^{\mathsf{T}}} \boldsymbol{\Omega}^{\mathrm{VO}} \mathbf{e}^{\mathrm{VO}} + \sum_{<i,j>\in \mathcal{C}^{\mathrm{HR}}} \mathbf{e}^{\mathrm{HR}^{\mathsf{T}}} \boldsymbol{\Omega}^{\mathrm{HR}} \mathbf{e}^{\mathrm{HR}} \tag{2.28}$$

$$\mathbf{x}^* = \arg\min_{\mathbf{x}} \sum_{<i,j>\in \mathcal{C}^{\mathrm{HR}}} \mathbf{e}^{\mathrm{HR}^{\mathsf{T}}} \boldsymbol{\Omega}^{\mathrm{HR}} \mathbf{e}^{\mathrm{HR}} + \sum_{<i,j>\in \mathcal{C}^{\mathrm{LM}}} \mathbf{e}^{\mathrm{LM}^{\mathsf{T}}} \boldsymbol{\Omega}^{\mathrm{LM}} \mathbf{e}^{\mathrm{LM}} + \sum_{<i,j>\in \mathcal{C}^{\mathrm{LC}}} \mathbf{e}^{\mathrm{LC}^{\mathsf{T}}} \boldsymbol{\Omega}^{\mathrm{LC}} \mathbf{e}^{\mathrm{LC}} \tag{2.29}$$

respectively, where $\mathcal{C}^{\mathrm{VO}}$, $\mathcal{C}^{\mathrm{HR}}$, $\mathcal{C}^{\mathrm{LM}}$, and $\mathcal{C}^{\mathrm{LC}}$ denote index sets containing measurements of VO, headings, landmark projections, and landmark locations; subscripts in error functions are omitted for convenient representation. In this work, each landmark vertex will be projected to a single camera pose vertex. Landmark locations are considered identical if a landmark is being tracked constantly, and we may set $\boldsymbol{\Omega}^{\mathrm{LC}}$ large to represent identical vertices.

By considering the non-convexity of the optimization problem, the initial pose estimates are set not to identity matrices but to VO-estimated poses such that the local minimum can be avoided. In TC-HRPE, since we are not estimating structure from motion, all feature vertices are fixed to avoid unnecessary optimization on 3D feature points.

2.3.3 Maintenance of the Dynamic Graph

The real-time performance of state estimation depends extensively on the graph size. In the SLAM problem, post-processing is acceptable to exploit all measurements, but it is not the case for VO. Although some approaches have been proposed for reducing computational load by minimizing graph size [Carlevaris-Bianco et al., 2014, Carlevaris-Bianco and Eustice, 2013], we implement the straightforward sliding window to restrain the unlimited growth of the computational load. During graph maintenance, all vertices and their edges out of the sliding window will be dropped out.

We have discussed the influence of vertex removal theoretically in Section 2.2.4. If all current vertices are optimal, the graph optimization problem can be simplified by deleting all edges that are not adjacent to the current vertex. However, a sliding window is essential in our proposed model because (i) vertices are never optimal with the existence of measurement noise, and (ii) landmark location measurements in TC-HRPE require the simultaneous existence of several pose vertices.

2.4 Simulation Studies

The KITTI odometry dataset has been selected to run simulative tests. Since no heading reference is available in the dataset, orientation measurements from an OxTS RT3003 Inertial and GPS Navigation System with manually added Gaussian noise $\mathcal{N}\left(\mathbf{0}, \text{diag}(\sigma_r^2, \sigma_p^2, \sigma_y^2)\right)$ for roll, pitch, and yaw are regarded as heading measurements. In all simulations and experiments, `libviso2` [Geiger et al., 2011a] and `g2o` [Kümmerle et al., 2011] are used for VO implementation and graph optimization, respectively.

Simulation results with sliding window size 10 and Euler angle noise parameters $\sigma_r = \sigma_p = \sigma_y = 5$ degrees are listed in Table 2.1; several sequences with relatively long traveling distances are selected during the evaluation to demonstrate the pose estimation results of the proposed approach with large rotational drifts. As this work considers pose estimation for ground vehicles, we focus on the translation error on the camera frame $x - z$ plane and rotation error on yaw.

Table 2.1 Simulation results of SVO, loosely coupled HRPE (LC-HRPE), and tightly coupled HRPE (TC-HRPE) in KITTI sequences 00, 02, 05, and 08. *Data format in translation error (m)/yaw error (deg).

Seq	Dist (m)	SVO		LC-HRPE		TC-HRPE	
		Avg Error*	Std Dev*	Avg Error*	Std Dev*	Avg Error*	Std Dev*
00	3,724	34.56/10.21	27.76/5.60	10.39/1.49	7.65/1.12	20.79/0.91	10.73/0.87
02	5,067	61.00/10.38	45.11/5.37	4.10/1.46	1.72/1.14	9.15/0.95	3.21/0.66
05	2,205	22.32/5.93	17.46/3.32	19.61/1.61	10.34/1.25	15.14/1.98	7.09/1.45
08	3,222	58.29/7.45	39.11/5.01	23.75/1.64	12.81/1.26	27.62/0.58	11.65/0.42

The results show that the proposed approach improves VO by reducing translational and rotational pose estimation errors and their standard deviations. In particular, yaw estimation errors for all evaluated sequences are shown in Figure 2.5. It is hard to fully compensate for drift error in VIO since IMU also suffers from inherent bias error [Bloesch et al., 2015]. Compared with VIO, which

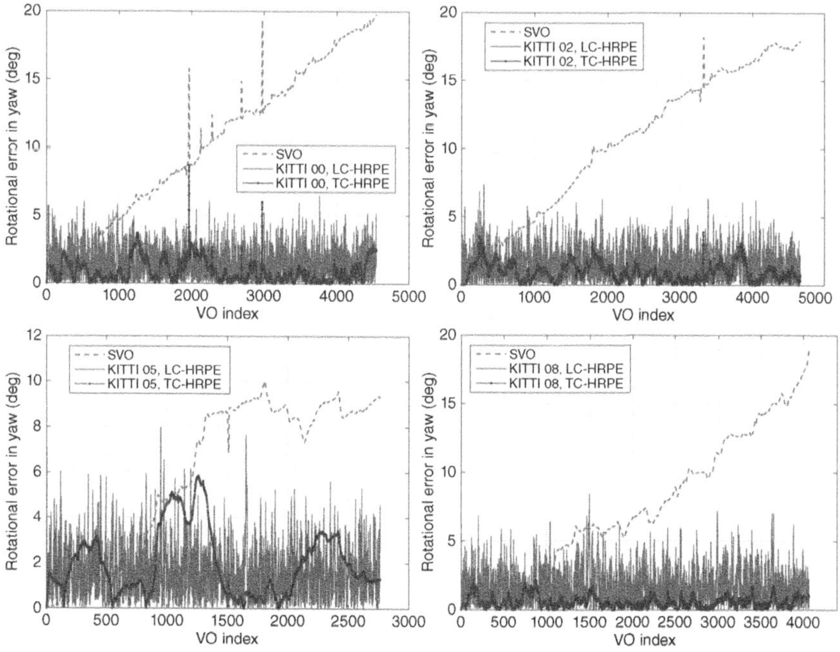

Figure 2.5 Yaw estimation error of KITTI sequences 00, 02, 05, and 08 with VO index, with heading measurement error $\sigma_r = \sigma_p = \sigma_y = 5$ degrees and sliding window size 10.

tends to be unbounded in estimation error [Usenko et al., 2016, Liu et al., 2016], it is evident that the rotation errors are bounded with the assistance of heading measurements. Moreover, we notice that TC-HRPE has more robust estimation performance on rotation due to its smaller error standard deviations.

The noise level of the heading measurements and the sliding window size play an essential role in algorithm performance. Next, we discuss accuracy and efficiency with various sliding window sizes and heading measurement errors.

2.4.1 Accuracy with Respect to Heading Measurement Error

For a particular heading reference sensor, since all measurements are represented in the sensor local frame, the system error caused by a misalignment between the sensors and the vehicle body frame has no effect on estimation accuracy. To investigate the influence of random error, Figure 2.6 shows the changing trend while the standard deviations of roll, pitch, and yaw measurement errors vary from 5 to 20 degrees. It is observed that accurate heading measurements generally lead to superior rotation estimation. The proposed estimation approach reduces rotation error compared to raw heading measurements with noise. In most cases, the standard deviation of the estimation error increases with inferior heading measurements.

2.4.2 Accuracy with Respect to Sliding Window Size

As shown in Figure 2.7, there is no significant difference observed with varying sliding window sizes in the simulation. As discussed in Section 2.2.4, dropping out previous vertices and their adjacent edges will not influence the current estimation in an ideal case. However, to ensure that the heading reference measurements take effect in the abstraction model and stabilize the graph in practice, size 10 is appropriate in this work.

For the tightly coupled model, increasing the sliding window size enables the possibility of considering more features that are observable in multiple frames. Furthermore, a relatively long sliding window may help retain more information such that it is possible to achieve global optimization and mapping when real-time estimation is not required.

2.4.3 Time Consumption with Respect to Sliding Window Size

The sliding window size will definitely influence efficiency while optimizing the graph. In this work, we explore the time consumed, and the results are shown in Figure 2.8. A maximum of two features are allowed in each bucket, with width and height 50×50 in VO. For the four KITTI sequences we evaluated, we compute the execution time by averaging the time per pose given different heading measurement noise with standard deviations 5, 10, 15, and 20 degrees.

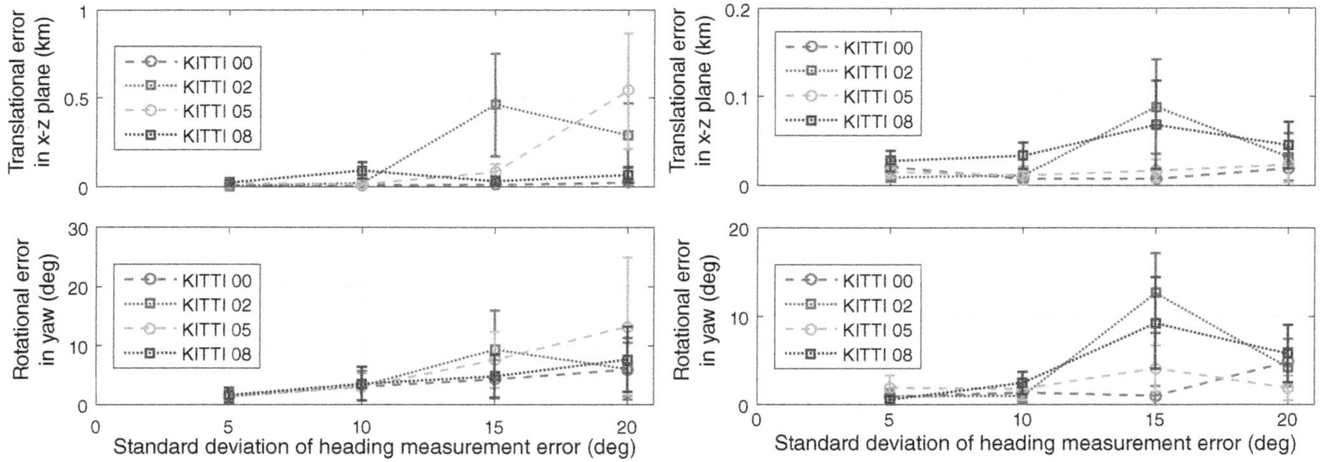

Figure 2.6 Estimation accuracy with heading measurement error using the loosely coupled (left) and tightly coupled (right) model, respectively.

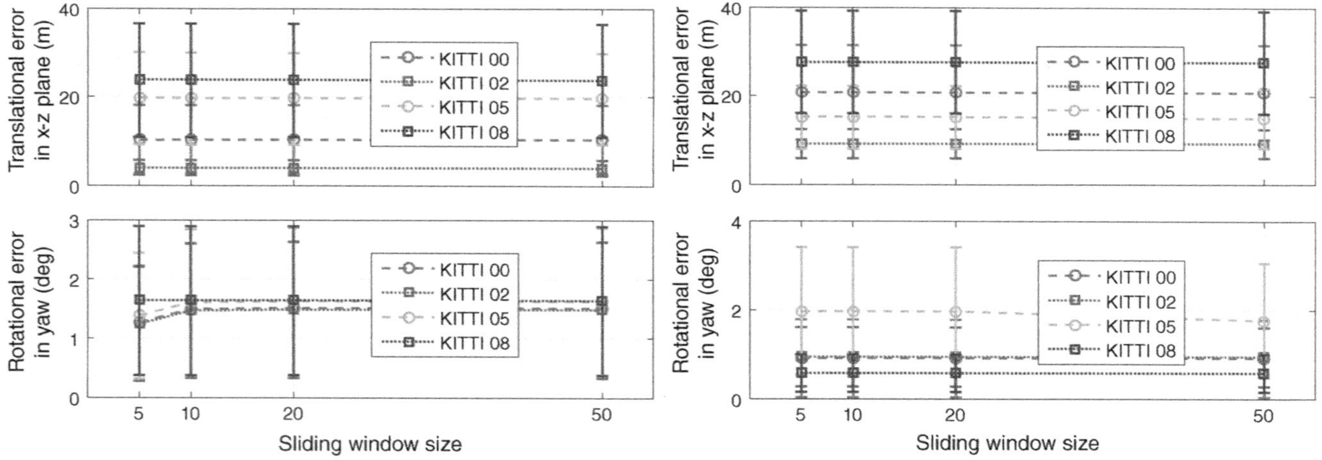

Figure 2.7 Estimation accuracy with respect to sliding window size using the loosely coupled (left) and tightly coupled (right) models with heading measurement error $\sigma_r = \sigma_p = \sigma_y = 5$ degrees.

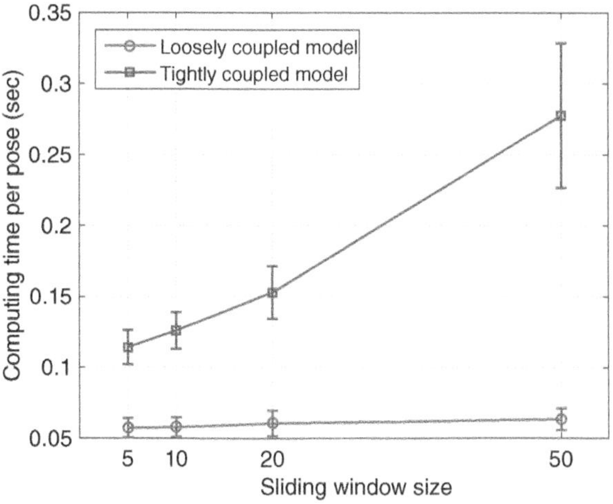

Figure 2.8 Average computing time per pose for simulated sequences.

The average execution time per pose for LC-HRPE and TC-HRPE is 0.06 second and 0.125 second with window size 10. Since features are detected and extracted for each new image frame in real time, the time for feature detection and extraction can be considered as constant complexity; thus, the time for optimization occupies a larger proportion of overall time consumption as the sliding window size becomes larger. By selecting an appropriate abstraction model and sliding window size according to the hardware performance, real-time implementation is attainable.

2.5 Experimental Results

To test the proposed approach further, we have conducted experimental studies using sequences we collected near the Nanyang Technological University (NTU) campus. The sequences contain challenging driving scenarios including passing humps, image intensity variations, and frequent acceleration/deceleration (see Figure 2.9 for graphical illustrations).

2.5.1 Experimental Platform

The experimental platform for data collection is shown in Figure 2.10. A stereo vision system consisting of two Flea3 image sensors (FL3-U3-13Y3M-C) and Kowa 1/2-inch lenses has been built to collect stereo images at a resolution of 640×422 with a baseline of 0.3 meter. Heading measurements in the NED frame are obtained from the DJI A3 GPS-IMU module, which measures the Earth's magnetic field and fuses it with internally measured linear acceleration and angular velocity. The stereo images are recorded at 40 frames per second

Figure 2.9 Outliers at turning (upper), passing humps (middle), and intensity variation (lower) in NTU sequences 01–04.

but have been down-sampled to 10 Hz. Fused translation poses from the DJI A3 GPS-IMU module are used as the vehicle's ground-truth position at 50 Hz. To evaluate the proposed approach, the ground truth is synchronized with the estimated pose using the ROS `Synchronizer` filter templated on `approximateTime` policy. The information matrices for VO (LC-HRPE), heading measurements (LC-HRPE), same feature constraint (TC-HRPE), feature-pose

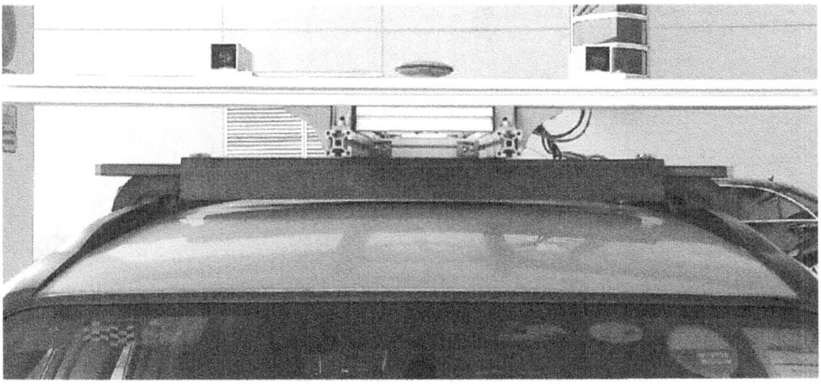

Figure 2.10 Our experimental platform.

constraint (TC-HRPE), and heading measurements (TC-HRPE) are $2000\mathbf{I}_6$, $100\mathbf{I}_3$, $10000\mathbf{I}_3$ $0.001\mathbf{I}_2$, and $2000\mathbf{I}_3$, respectively.

2.5.2 Pose Estimation Performance

We only evaluate translation error quantitatively because orientations have been used as heading measurements and thus cannot be taken as fair ground truth. The qualitative performance of rotation estimation is still accessible from the vehicle trajectories. The pose estimation performance for LC-HRPE and TC-HRPE is shown quantitatively in Table 2.2, where the results demonstrate that the translation estimation error has been substantially reduced with the assistance of heading measurements. Specifically, the average translation errors for SVO, LC-HRPE, and TC-HRPE are 153.85 meters, 24.29 meters, and 23.80 meters. The proposed approach has been improved the robustness of estimation according to the smaller error standard deviation.

Table 2.2 Experimental results of SVO, loosely coupled HRPE (LC-HRPE), and tightly coupled HRPE (TC-HRPE) in NTU sequences 01–04. Frequency in Hz.

Seq	Dist (m)	SVO: Trans(m)			LC-HRPE: Trans(m)			TC-HRPE: Trans(m)		
		Avg err	Std dev	Freq	Avg err	Std dev	Freq	Avg err	Std dev	Freq
01	1,244	132.32	115.48	22.6	22.3	10.62	20.8	14.85	10.85	11.5
02	1,249	184.57	120.00	20.3	19.53	8.13	19.0	35.62	9.41	10.8
03	860	163.44	122.68	20.3	21.41	8.19	19.7	28.20	12.24	10.9
04	1,204	141.98	107.64	19.9	31.96	26.76	19.4	19.30	14.62	11.2
Tot	4,557	153.85	-	20.8	24.29	-	19.7	23.80	-	11.1

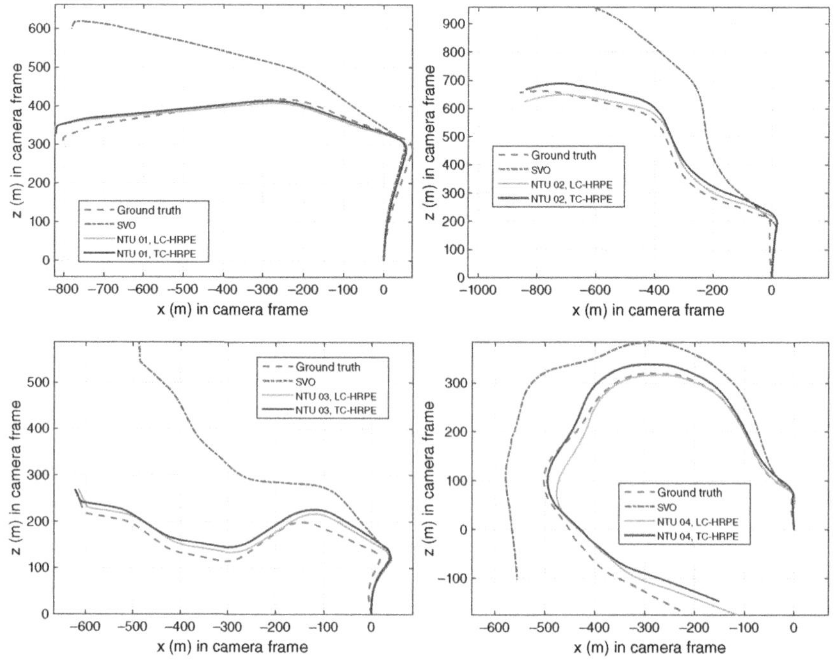

Figure 2.11 Experimental results based on self-collected sequences NTU 01–04.

Figure 2.11 and Figure 2.12 illustrate the comparative results on vehicle trajectories and translation estimation errors for all four sequences. Rotation error increases rapidly, especially when the vehicle is making turns in environments with outliers, illumination variation, and poor features. For the same reason, translation error tends to be unbounded with SVO only. The proposed approach slows the translation error growth rate such that the position is still effective in the case of a large rotation error in VO. Although it is extremely difficult to bound translation error due to egomotion iteration and error accumulation, accurate rotation estimation contributes significantly to VO performance.

2.5.3 Real-Time Performance

The proposed system has been evaluated on a desktop computer with an Intel Core i7-4770 CPU 3.40GHz and 12GB of memory. Experimental real-time performance is listed in Table 2.2, which shows that the average execution frequencies for SVO, LC-HRPE, and TC-HRPE are 20.8 Hz, 19.7 Hz, and 11.1 Hz, respectively. The loosely coupled model brings a little extra computational load to pure odometry compared to the tightly coupled model that doubles operation time. For TC-HRPE, not only the sliding window size but also the feature detection and extraction parameters significantly influence real-time performance. In our

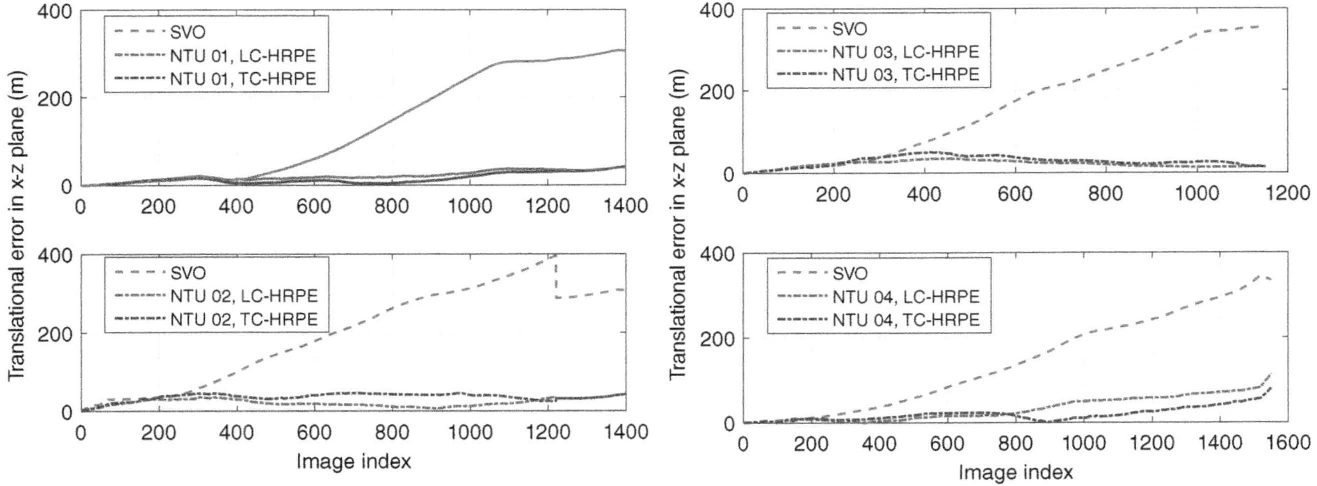

Figure 2.12 Translation estimation error based on NTU 01–04.

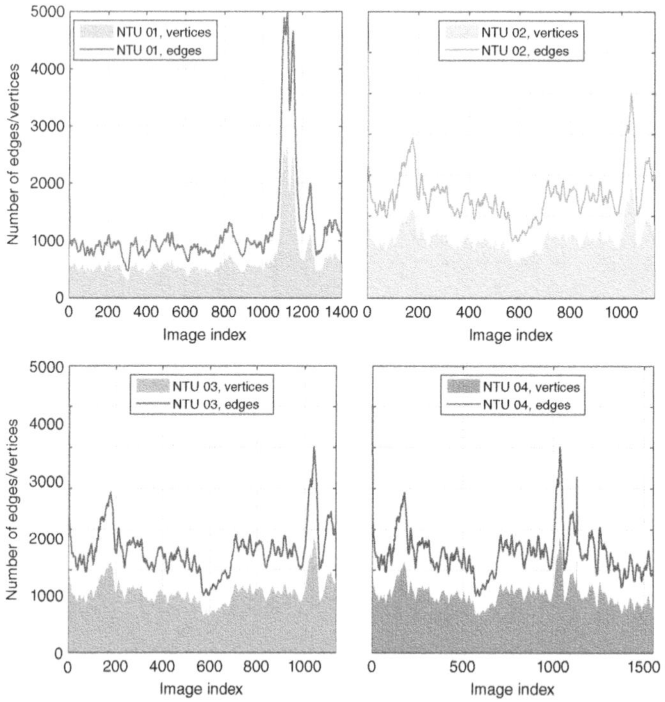

Figure 2.13 The sizes of graphs to be optimized based on the tightly coupled model and self-collected sequences NTU 01–04.

experiments, the average graph edge number at each iteration for TC-HRPE can be found in Figure 2.13. Reducing the feature detection and selection threshold may help increase VO performance but will lead to delayed results.

2.6 Conclusion

In this chapter, we have presented a stereo pose estimation framework that needs assistance from heading references, with elaboration on the proposed abstraction model and evaluations using both public and self-collected datasets. The proposed system has demonstrated its effectiveness and robustness by significantly reducing translation and rotation estimation error compared to pure stereo visual odometry. The extra computational cost is acceptable for the real-time requirement of pose estimation. The absolute heading reference bounds the rotation drift, whereas the translation drift will increase unboundedly at a significantly lower rate. As with all other visual pose estimation approaches, poor illumination conditions and outliers in image features will cause performance degradation.

3

Road-Constrained Localization Using Cloud Models

The problem of autonomous vehicle localization has been well-studied over the last few years [Castellanos and Tardos, 2012]. Although the global positioning system (GPS)-based approach has become one of the most popular positioning solutions, it is necessary to develop non-GPS-based approaches due to inherent drawbacks of GPS, such as instability in urban canyon environments [Cui and Ge, 2003] and poor accuracy at short distances [Kaplan and Hegarty, 2005]. The camera, which is a cost-effective, easy-to-use, and lightweight sensor, has been widely used to solve the localization problem, and impressive results have been achieved [Milford et al., 2014, Lim et al., 2015].

Visual odometry (VO), the most commonly used method, is a self-contained dead-reckoning localization process and has been successfully implemented in Mars Exploration Rovers [Maimone et al., 2007]. As discussed in Chapter 2, it is known that VO suffers from drift, which is the primary factor limiting its long-range applications. Compared to stereo visual odometry (SVO), monocular visual odometry (MVO) shows advantages in cost and size. However, scale ambiguity makes MVO even harder for pose recovery without adding constraints [Kitt et al., 2011]. Map-assisted positioning benefits from precise geographical information and provides a new perspective for vehicle localization, especially in urban areas. With the assumption that the vehicle is always on the road, the geometric shapes of road networks can be used as constraints to assist with position estimation [Brubaker et al., 2013, Floros et al., 2013, Merriaux et al., 2015].

The objective of this chapter is to localize a vehicle equipped with a single camera and a freely available map in the global map frame. To complete this goal, a framework is presented that combines MVO measurement and a map. In this framework, MVO measurement, i.e. the incremental translation between two consecutive image frames, is used to generate particles (possible trajectories)

Multimodal Perception and Secure State Estimation for Robotic Mobility Platforms, First Edition.
Xinghua Liu, Rui Jiang, Badong Chen, and Shuzhi Sam Ge.

under the newly proposed uniform-Gaussian distribution; then, shape matching between trajectories and the map is applied to weight those particles.

The rest of this chapter proceeds as follows: Section 3.1 introduces cloud models and presents definitions and properties of two clouds: uniform-Gaussian distribution and Gaussian-Gaussian distribution. Section 3.2 describes the localization framework, where the uniform-Gaussian probability model, map-assisted shape matching, and particle resampling are incorporated in a unified localization framework. Extension to the Gaussian-Gaussian model is also discussed. In Sections 3.3 and 3.4, the proposed approach is implemented with SVO and MVO, and experimental results are compared and discussed. Final remarks for this chapter are given in Section 3.5.

3.1 Preliminaries

3.1.1 Scaled Measurement Equations for Visual Odometry

It is impossible to obtain measurements without uncertainty. The conventional Gaussian probability assumption makes sense in dealing with MVO measurement randomness (without considering the scaling factor). However, MVO cannot recover a trajectory's absolute scale since scale ambiguity exists. This section demonstrates measurement equations for MVO and makes an assumption in preparation for modeling measurement uncertainties.

VO outputs a raw transformation matrix by measuring camera pose differences between consecutive image frames, as shown previously in Section 2.1.1. Since maps are two-dimensional projections from real geographic information, to incorporate road constraints into VO measurements, two-dimensional simplification is necessary. Moreover, as the focus of the chapter is translation rather than rotation in the localization problem, the relation between raw measurements and scaled measurements (VO aligned with actual scale) can be expressed as

$$\mathbf{t}_{k,k-1} = s_k \mathbf{t}_{k,k-1}^{\text{raw}} \tag{3.1}$$

where $s_k \in \mathbb{R}^+$ is a positive scaling factor at time instant k. Equation (3.1) is more generalized and can be applied to SVO by letting $s_k = 1$. To push forward, we make the following assumption:

Assumption In the MVO measurement model Eq. (3.1), for every time instant k, s_k is independent with each element in $\mathbf{t}_{k,k-1}$, and the elements in $\mathbf{t}_{k,k-1}$ are also mutually independent. □

3.1.2 Cloud Models

The cloud model was first proposed by Li et al. (2007, 2009) to convert between qualitative and quantitative concepts. In the Gaussian cloud model discussed in Li and Du (2007), three numerical characteristics (expected value, entropy, and hyper-entropy) are defined to describe a qualitative concept. By incorporating randomness and fuzziness into unified consideration, the cloud model can represent concepts with uncertainty, where each cloud drop contributes to the concept with a particular "degree of contribution."

Definition 3.1 ***(Cloud [Li et al., 2009])*** Let U be a universal set described by a precise number, and C be the qualitative concept related to U. If there is a number $x \in U$ that randomly realizes the concept C, and the certainty degree of x for C, i.e. $\mu(x) \in [0,1]$, is a random value with stable tendency:

$$\mu : U \to [0,1] \quad \forall x \in U \quad x \to \mu(x) \tag{3.2}$$

Then the distribution of x on U is defined as a **cloud** $C(x)$, and every x is defined as a **cloud drop**. The certainty degree μ also refers to a membership grade.

The cloud model has shown promising results in modeling uncertainty. In particular, it has been successfully implemented in decision-making [Wang et al., 2015], situation prediction [Hu and Qiao, 2016], path planning [Sun et al., 2015], and stochastic optimization [Kavousi-Fard et al., 2016]. Although the Gaussian cloud model is the most-studied cloud model, due to its universality, different types of clouds can be created according to various probability distributions, such as uniform clouds, power-law clouds, trapezium clouds, etc. The appropriate choice of a cloud model is based on accurate comprehension of the physical process that describes the designated problem. In this chapter, two types of cloud-inspired probabilistic distributions are presented based on VO measurement principles such that uncertainties in VO can be represented.

3.1.3 Uniform Gaussian Distribution (UGD)

Inspired by the cloud model, the uniform Gaussian distribution (UGD) is proposed to represent MVO measurements with respect to drift and scale ambiguity. With Eq. (3.1), we use a Gaussian-distributed random vector $\mathbf{X}$ to represent raw translation measurement $\mathbf{t}^{\text{raw}}_{k,k-1}$. Since we have no a priori knowledge about the scaling factor s_k, it is reasonable to assume that s_k can be modeled using a random variable S that obeys a uniform distribution in the interval $[a, b]$, where $[a, b]$ denotes the current probable range of s_k. Thus, the scaled translation measurement $\mathbf{t}_{k,k-1}$

is expressed as a random vector $\mathbf{Y} = S\mathbf{X}$, which leads to the definition of UGD as follows:

Definition 3.2 ***(Uniform Gaussian Distribution, UGD)*** Given random variable S and random vector $\mathbf{X}$, the variate $\mathbf{Y} = S\mathbf{X}$ obeys UGD $UG(a, b, E\mathbf{X}, \boldsymbol{\Sigma})$ if $S \sim U(a, b)$ and $\mathbf{X} \sim N(E\mathbf{X}, \boldsymbol{\Sigma})$, where $U(\cdot)$ and $N(\cdot)$ are abbreviations for uniform and Gaussian distribution, respectively.

This definition generalizes a Gaussian distribution: if $a = b$, then S will be a constant, which results $\mathbf{Y}$ in a multivariate bell-shaped distribution. Based on Assumption 3.1, S is independent with each variable in $\mathbf{X}$, and variables in $\mathbf{X}$ are mutually independent. Several important properties, some of which will be used for parameter estimation, are derived here.

Property 3.1 ***(PDF of UGD)*** The probability density function (PDF) of the jth random variable in $\mathbf{Y}$ can be represented as [Rohatgi and Saleh, 2015]

$$f_{Y_j}(y) = \int_{-\infty}^{+\infty} f_S(s) f_{X_j}\left(\frac{y}{s}\right) \frac{1}{|s|} \mathrm{d}s \tag{3.3}$$

If $0 < a \leq b$, this PDF can be simplified as

$$f_{Y_j}(y) = \frac{1}{(b-a)\sigma_j\sqrt{2\pi}} \int_a^b \frac{1}{s} \exp\left\{-\frac{\left[y/s - E(X_j)\right]^2}{2\sigma_j^2}\right\} \mathrm{d}s \tag{3.4}$$

where σ_j^2 denotes the jth diagonal element in covariance matrix $\boldsymbol{\Sigma}$.

Property 3.2 ***(Expectation of UGD)*** The expectation of $\mathbf{Y}$ is the multiplication of the expectations of S and $\mathbf{X}$:

$$E\mathbf{Y} = \frac{a+b}{2} E\mathbf{X} \tag{3.5}$$

Property 3.3 ***(Variance of UGD)*** The variance of the jth random variable in $\mathbf{Y}$ can be represented as

$$\begin{aligned} D(Y_j) &= D(X_j)E(S)^2 + D(S)E(X_j)^2 + D(S)D(X_j) \\ &= \frac{a^2+b^2-2ab}{12} E(X_j)^2 + \frac{a^2+b^2+ab}{3} \sigma_j^2 \end{aligned} \tag{3.6}$$

Proof: Define

$$f_{\text{norm}}(x) = \frac{1}{\sigma_j\sqrt{2\pi}} \exp\left\{-\frac{\left[x - E(X_j)\right]^2}{2\sigma_j^2}\right\}$$

Then we have

$$D(Y_j) = \int_{-\infty}^{+\infty} [y - E(Y_j)]^2 f_{Y_j}(y)\mathrm{d}y$$

$$= \int_{-\infty}^{+\infty} [y - E(Y_j)]^2 \frac{1}{(b-a)\sigma_j\sqrt{2\pi}} \int_a^b \frac{1}{s} \exp\left\{-\frac{[y/s - E(X_j)]^2}{2\sigma_j^2}\right\} \mathrm{d}s\mathrm{d}y$$

$$= \frac{1}{(b-a)} \int_a^b \frac{1}{s} \int_{-\infty}^{+\infty} [sx - E(S)E(X_j)]^2 f_{\text{norm}}(x)\mathrm{d}(sx)\mathrm{d}s$$

$$= \frac{1}{(b-a)} \int_a^b \left\{ s^2 \int_{-\infty}^{+\infty} x^2 f_{\text{norm}}(x)\mathrm{d}x - 2E(S)E(X_j)s \int_{-\infty}^{+\infty} x f_{\text{norm}}(x)\mathrm{d}x \right.$$

$$\left. + E^2(S)E^2(X_j) \int_{-\infty}^{+\infty} f_{\text{norm}}(x)\mathrm{d}x \right\} \mathrm{d}s$$

$$= \frac{1}{(b-a)} \int_a^b \left\{ s^2 \left[E^2(X_j) + \sigma_j^2\right] - 2E(S)E^2(X_j)s + E^2(S)E^2(X_j) \right\} \mathrm{d}s$$

$$= \frac{a^2 + b^2 - 2ab}{12} E^2(X_j) + \frac{a^2 + b^2 + ab}{3} \sigma_j^2$$

□

Property 3.4 ***(Third Central Moment of UGD)*** The third central moment of the jth random variable in **Y** can be expressed as

$$S(Y_j) = \frac{a^3 - ab^2 - a^2b + b^3}{4} E(X_j)\sigma_j^2 \tag{3.7}$$

Proof: Similar to the proof of Property 3.3, define

$$f_{\text{norm}}(x) = \frac{1}{\sigma_j\sqrt{2\pi}} \exp\left\{-\frac{[x - E(X_j)]^2}{2\sigma_j^2}\right\}$$

and given the third non-central moment of the jth random variable in **X** as $E(X_j)^3 + 3E(X_j)\sigma_j^2$, we have

$$S(Y_j) = \int_{-\infty}^{+\infty} [y - E(Y_j)]^3 f_{Y_j}(y)\mathrm{d}y$$

$$= \int_{-\infty}^{+\infty} [y - E(Y_j)]^3 \frac{1}{(b-a)\sigma_j\sqrt{2\pi}} \int_a^b \frac{1}{s} \exp\left\{-\frac{[y/s - E(X_j)]^2}{2\sigma_j^2}\right\} \mathrm{d}s\mathrm{d}y$$

$$= \frac{1}{(b-a)} \int_a^b \frac{1}{s} \int_{-\infty}^{+\infty} [sx - E(S)E(X_j)]^3 f_{\text{norm}}(x)\mathrm{d}(sx)\mathrm{d}s$$

$$= \frac{a^3 - ab^2 - a^2b + b^3}{4} E(X_j)\sigma_j^2$$

□

Algorithm 1 UGD sampling

Input: UGD parameters $(a, b, E\mathbf{X}, \mathbf{\Sigma})$, number of samples N.

Output: Samples $\mathbf{y}_i, i \in \{1, \dots, N\}$

1: **for** $i \leftarrow 1$ to N **do**

2: $s_i \leftarrow \mathtt{Uniform}(a, b)$ // generating uniform-distributed s_i

3: $\mathbf{x}_i \leftarrow \mathtt{Norm}(E\mathbf{X}, \mathbf{\Sigma})$ // generating Gaussian-distributed $\mathbf{x}_i$

4: $\mathbf{y}_i \leftarrow s_i\mathbf{x}_i$

5: **end for**

Given the parameters of UGD, samples can be generated according to Algorithm 1.

3.1.4 Gaussian-Gaussian Distribution (GGD)

To allow more flexibility in modeling VO measurements, we further define the Gaussian-Gaussian distribution (GGD). Its properties can be obtained similarly to those of the UGD, as follows:

Definition 3.3 ***(Gaussian-Gaussian Distribution, GGD)*** Let U be the universe of discourse and GGD be a qualitative concept in U. If $\mathbf{Y} = \mathbf{S} \circ \mathbf{X} \in U$ is a random instantiation of concept GGD, where $\circ$ denotes the Hadamard product (or the Schur product); $\mathbf{S}$ and $\mathbf{X}$ are random vectors that obey independent Gaussian distribution $N(E\mathbf{S}, \Sigma_S)$ and $N(E\mathbf{X}, \Sigma_X)$, respectively; and the certainty degree of $\mathbf{Y}$ belonging to concept GGD satisfies where $\mathrm{diag}(\mathbf{S}^T)$

$$\mu(\mathbf{Y}) = \exp\left\{-\frac{1}{2}\left[\mathrm{diag}(\mathbf{S}^T)^{-1}\mathbf{Y} - E\left(\mathbf{X}\right)\right]^T \Sigma_X^{-1} \left[\mathrm{diag}(\mathbf{S}^T)^{-1}\mathbf{Y} - E\left(\mathbf{X}\right)\right]\right\} \tag{3.8}$$

denotes the diagonal matrix with corresponding diagonal entries from $\mathbf{S}^T$, then the distribution of $\mathbf{Y}$ in the universe U is a multivariate Gaussian-Gaussian distribution, which can be denoted by $\mathbf{Y} \sim GGD(E\mathbf{S}, \Sigma_S, E\mathbf{X}, \Sigma_X)$.

$$f_{Y_j}(y) = \int_{-\infty}^{+\infty} f_{S_j}(s) f_{X_j}\left(\frac{y}{s}\right) \frac{1}{|s|} \mathrm{d}s \tag{3.9}$$

$$= \frac{1}{2\sigma_{S_j}\sigma_{X_j}\pi} \int_{-\infty}^{+\infty} \exp\left\{-\frac{\left[s - E(S_j)\right]^2}{2\sigma_{S_j}^2}\right\} \exp\left\{-\frac{\left[y/s - E(X_j)\right]^2}{2\sigma_{X_j}^2}\right\} \frac{1}{|s|} \mathrm{d}s \tag{3.10}$$

We now move on and discuss several statistical characteristics of GGD, including probability density functions and moments. In the following discussion, we assume $\mathbf{Y} \sim GGD(E\mathbf{S}, \Sigma_S, E\mathbf{X}, \Sigma_X)$, and we use Y_j to represent the jth random variable in random vector $\mathbf{Y}$ for the sake of analysis. These properties will be used for parameter estimation.

Property 3.5 ***(PDF of GGD)*** The probability density function (PDF) of Y_j can be represented as where $\sigma_{S_j}^2$ and $\sigma_{X_j}^2$ denote the jth diagonal element in covariance matrices Σ_S and Σ_X, respectively.

Property 3.6 ***(Expectation)*** The expectation or first moment of Y_j and $\mathbf{Y}$ can be obtained from

$$E(Y_j) = E(S_j)E(X_j) \tag{3.11}$$

$$E\mathbf{Y} = E\mathbf{S} \circ E\mathbf{X} \tag{3.12}$$

Property 3.7 ***(Variance)*** The variance or second central moment of Y_j can be obtained from

$$D(Y_j) = D(X_j)E(S_j)^2 + D(S_j)E(X_j)^2 + D(S_j)D(X_j) \tag{3.13}$$

$$= \sigma_{X_j}^2 E(S_j) + \sigma_{S_j}^2 E(X_j) + \sigma_{S_j}^2 \sigma_{X_j}^2 \tag{3.14}$$

3.2 Map-Assisted Ground Vehicle Localization

The framework of the proposed approach is shown in Figure 3.1. The map preparation process requires a rough positioning result such that a particular region of

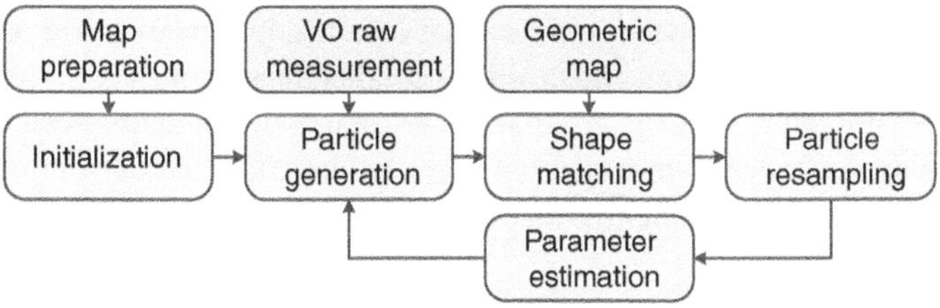

Figure 3.1 Framework of the proposed map-assisted localization approach.

the map can be prepared. If the rough position is not available, any other position indication, such as a block name, will do, as long as a rough area can be identified. The duration of the initialization process depends on the practical road conditions. Since only a sufficiently complex trajectory can be used for shape matching, the main loop will not be executed unless the trajectory satisfies certain geometric conditions.

Each time particles are generated based on the UGD, shape matching is applied to gauge the similarity between the map and every probable trajectory that contains historical and current positions. Particle *saliency* and *consistency* are proposed to indicate the representativeness of a particle for VO measurement and the degree of similarity between the estimated trajectory and the map, respectively. Then particle weights are determined by considering both *saliency* and *consistency*. Next, a resampling process is carried out to select particles with higher weights and discard particles with lower weights. The remaining particles are used to update the parameters of the UGD. Using the framework mentioned earlier, the vehicle is localized while VO measurement uncertainties are reduced simultaneously.

3.2.1 Measurement Representation with UGD

Scaled measurements of MVO can be regarded as uniform Gaussian distributed samples $\mathbf{y}_i$ that are the multiplications of uniformly distributed samples s_i and Gaussian distributed samples $\mathbf{x}_i$. Since only a single measurement can be obtained at a certain time, in the meaning of maximum likelihood estimation, the raw measurement is estimated as the expectation of Gaussian distribution $E\mathbf{X}$. The covariance matrix $\boldsymbol{\Sigma}$, which represents the uncertainty of a VO result, can be obtained from the exponential drift model [Jiang et al., 2010] as $\boldsymbol{\Sigma} = \gamma_0^{d_{\text{VO}}}\boldsymbol{\Sigma}_0$, where γ_0 and $\boldsymbol{\Sigma}_0$ are platform-dependent constants denoting the base growth rate and the initial covariance matrix, respectively; d_{VO} is the distance measured by VO. Given the initial value, subsequent parameters a and b can be obtained iteratively based on the parameter estimation algorithm, which will be detailed in Section 3.2.3.

Given a raw measurement, a group of samples is created to represent the scaled MVO measurement. In order to measure the representativeness of any sample with respect to the raw measurement, *saliency* $\mu_{\text{VO}}(\mathbf{y}_i) \in (0,1]$ can be proposed uniquely for each sample as

$$\mu_{\text{VO}}(\mathbf{y}_i) = \exp\left\{-\frac{1}{2}\left[\frac{\mathbf{y}_i}{s_i} - E\left(\frac{\mathbf{y}_i}{s_i}\right)\right]^{\mathrm{T}} \boldsymbol{\Sigma}^{-1} \left[\frac{\mathbf{y}_i}{s_i} - E\left(\frac{\mathbf{y}_i}{s_i}\right)\right]\right\} \tag{3.15}$$

$$= \exp\left\{-\frac{1}{2}\left[\mathbf{x}_i - E\mathbf{X}\right]^{\mathrm{T}} \boldsymbol{\Sigma}^{-1} \left[\mathbf{x}_i - E\mathbf{X}\right]\right\} \tag{3.16}$$

This definition is rational since the sample that satisfies $\mathbf{x}_i = E\mathbf{X}$ is regarded as the most representative sample for VO measurement. In practice, the scaled measurement from VO can be represented as a particle set, where each particle contains a sample and its corresponding *saliency*.

Given the raw trajectory from MVO, the concept of a particle can be generalized to represent a scaled trajectory containing multiple samples from MVO raw measurements at different times and their (identical) saliency under the same scaling factor.

3.2.2 Shape Matching Between Map and Particles

The road shape information in the map is used to filter out trajectories with inaccurate MVO scales. This idea is quite similar to shape matching in computer vision, whose goal is to find the best alignment between two edge maps [Jiang et al., 2009, Joo et al., 2009]. To use the map effectively, we first convert it to a binary edge map $\mathcal{U} = \{\mathbf{u}_p\}$, where all the street segments in the map are discretized and represented in pixels. Then each particle $\mathbf{y}_i$ representing a possible trajectory is plotted and converted to an edge map $\mathcal{V}_i = \{\mathbf{v}_q\}$. The shape matching distance between $\mathcal{U}$ and $\mathcal{V}_i$ is given by the average distance between each pixel $\mathbf{v}_q \in \mathcal{V}_i$ and its nearest edge in $\mathcal{U}$ [Barrow et al., 1977]

$$d_{\mathrm{SM}}(\mathcal{V}_i, \mathcal{U}) = \frac{1}{n_{\mathrm{SM}}} \sum_{\mathbf{v}_q \in \mathcal{V}_i} \min_{\mathbf{u}_p \in \mathcal{U}} \|\mathbf{v}_q - \mathbf{u}_p\| \tag{3.17}$$

where $n_{\mathrm{SM}} = |\mathcal{V}_i|$ denotes the number of points in $\mathcal{V}_i$. In short, a shape-matching distance d_{SM} for each particle is computed according to a template-matching technique. Particles that fit the road shape well are assigned a small shape-matching distance, while particles inconsistent with the road constraints are given a large shape-matching distance.

To measure the degree to which the particle $\mathbf{y}_i$ is in accordance with vehicle's actual trajectory, the *consistency* $\mu_{\mathrm{SM}}(\mathbf{y}_i)$ is proposed. Since it is obvious that the *consistency* looks negatively associated with the shape-matching distance, some would assume that the ideal distribution of $d_{\mathrm{SM}}(\mathcal{V}_i, \mathcal{U})$ follows an exponential model [Floros et al., 2013]. However, as all roads on the map are generally modeled with the road center line, there is an inevitable offset between the vehicle's actual trajectory and the roads represented on the map. Based on this fact, μ_{SM} achieves the maximum when $d_{\mathrm{SM}}(\mathcal{V}_i, \mathcal{U}) > 0$.

In this work, the Chi-squared distribution has been used to approximate the relation between shape-matching distance and *consistency*. Thus, we write *consistency* as

$$\mu_{\mathrm{SM}}(\mathbf{y}_i) = \chi(d_{\mathrm{SM}}(\mathcal{V}_i, \mathcal{U}), k) \tag{3.18}$$

where $\chi(d_{\mathrm{SM}}(\mathcal{V}_i, \mathcal{U}), k)$ denotes the Chi-squared distribution with the degrees of freedom $k = 3$, and $d_{\mathrm{SM}}(\mathcal{V}_i, \mathcal{U})$ is the shape-matching distance for particle $\mathbf{y}_i$.

3.2.3 Particle Resampling and Parameter Estimation

The weight of each particle before normalization can be obtained from

$$\tilde{w}(\mathbf{y}_i) = \mu_{\mathrm{VO}}(\mathbf{y}_i)\mu_{\mathrm{SM}}(\mathbf{y}_i) \tag{3.19}$$

Herein, N particles with top weights are selected for location and parameter estimation. After resampling, weights need to be normalized for selected particles, and the normalization rule can be written as $w(\mathbf{y}_i) = \frac{\tilde{w}(\mathbf{y}_i)}{\sum_{i=1}^{N} \tilde{w}(\mathbf{y}_i)}$.

Estimating all parameters in UGD leads to higher-degree polynomial equations with no algebraic solution. In this particular resampling step, as the raw VO result is quite accurate over a short distance, the scaling factor plays a major role in selecting particles. To simplify the problem, we assume that the expectation and variance of samples remain the same after resampling. In other words, shape matching filters out particles with improper scaling factors, and the distribution of raw MVO measurements does not change much after the resampling process. The simplified parameter estimation problem can be formulated as follows: Given N samples $\mathbf{y}_i (i = 1, \cdots, N)$, estimate UGD parameters a, b with known parameters $\mathbf{EX}$ and $\mathbf{\Sigma}$.

According to Property 3.2–3.3, by using the method of moments, the parameters can be estimated based on the following equations

$$M_{1j} = \frac{\hat{a} + \hat{b}}{2} EX_j \tag{3.20}$$

$$M_{2j} = \frac{\hat{a}^2 + \hat{b}^2 - 2\hat{a}\hat{b}}{12} E(X_j)^2 + \frac{\hat{a}^2 + \hat{b}^2 + \hat{a}\hat{b}}{3} \sigma_j^2 \tag{3.21}$$

where $M_{1j} = \frac{1}{N} \sum_{i=1}^{N} y_{ij}$ and $M_{2j} = \frac{1}{N-1} \sum_{i=1}^{N} (y_{ij} - M_{1j})^2$ denote the expectation and variance of sample particles, respectively. The estimation results can be expressed as

$$\hat{a}_j = \frac{M_{1j}}{EX_j} - \sqrt{3 \frac{M_{2j} - \frac{\sigma_j^2}{EX_j^2} M_{1j}^2}{EX_j^2 + \sigma_j^2}} \tag{3.22}$$

$$\hat{b}_j = \frac{M_{1j}}{EX_j} + \sqrt{3 \frac{M_{2j} - \frac{\sigma_j^2}{EX_j^2} M_{1j}^2}{EX_j^2 + \sigma_j^2}} \tag{3.23}$$

where the estimated scale range $[\hat{a}, \hat{b}]$ is used to generate particles obeying $UG(\hat{a}, \hat{b}, \mathbf{EX}, \mathbf{\Sigma})$ at the next iteration.

3.2.4 Framework Extension to Other Cloud Models

The previously described framework can be extended by replacing the representation of VO measurement with other types of clouds, such as GGD, according to the assumed uncertainties, including drift and scale ambiguity. The detailed implementation using GGD is omitted here. Interested readers may refer to Yang et al. (2017b) for more information. The next two sections cover experimental results using UGD and GGD models.

3.3 Experimental Validation on UGD

Two sets of experiments have been conducted, and the results are shown to demonstrate the performance of the proposed approach. Although the main contribution of this work is intended to reduce MVO measurement uncertainties, including both drift and scale ambiguity, evaluation with SVO data is conducted first because localization with SVO can be regarded as a particular case with MVO when $a = b = 1$ is known. The SVO case with a fixed and known scaling factor makes it convenient for algorithm assessment.[1]

In the second set of experiments, we implement the proposed approach with MVO raw measurements. All experiments are conducted on the challenging KITTI dataset. Some sequences in KITTI have a very short or straight driving trajectory, making the vehicle difficult to localize. The "fundamental ambiguities" problem is also discussed in Brubaker et al. (2013a).

3.3.1 Configurations

In all experiments, OpenStreetMaps (OSMs) have been used as primitive maps. In the beginning, a rough positioning result from GPS is needed for map downloading. Then the map is revised by preserving drivable roads only. Before a VO raw measurement is used for particle generation, an initialization step is required. As the shape matching needs a trajectory with enough geometric characteristics, the first batch of particles will not be generated until a minimum distance $d_{\text{min}} = 300$ meters and a turning angle $\alpha_{\text{min}} = \frac{\pi}{4}$ is reached. Then, 2000 particles are generated. The map sizes for sequences 00, 02, 05, 08, and 09 are 725×836, 1159×1165, 730×583, 794×915, and 854×820 meters, respectively. A low-variance sampler is implemented in particle resampling. Parameters γ_0 and $\mathbf{\Sigma}_0$ are set to $\gamma_0 = 1.02$, $\mathbf{\Sigma}_0 = \text{diag}(2,2)$ for SVO and $\gamma_0 = 1.005$, $\mathbf{\Sigma}_0 = \text{diag}(0.1, 0.1)$ for MVO.

---

1 It is noted that the scale for SVO may be slightly off the theoretical value due to calibration error and feature selection/matching error.

Table 3.1 Quantitative comparison between the proposed approach and SVO localization in KITTI sequences 00, 02, 05, 08, and 09. (Distance in kilometers; mean and standard deviation of error in meters.)

		Error (proposed)			Error (SVO)			Error in Brubaker et al. (2013a)
Seq	**Dist**	**Avg**	**Std**	**Max**	**Avg**	**Std**	**Max**	**Avg**
00	3.72	3.14	1.16	10.28	37.11	29.63	107.23	2.1
02	5.06	11.46	7.47	25.42	66.08	47.37	172.71	4.1
05	2.20	4.14	2.12	8.89	14.40	13.30	45.60	2.6
08	3.21	5.01	3.01	12.65	34.70	28.12	98.20	2.4
09	1.70	5.75	2.29	11.84	16.63	34.91	9.72	4.2
Total	15.89	6.58	–	–	40.51	–	–	3.19

During the experiments, most of the computation time is reserved for shape matching and particle resampling. With C++ implementation, it takes 1.5 ms per particle for shape matching and particle resampling on a mobile workstation with an i7-4710MQ processor. Since particle resampling at every time instant is unnecessary, it is promising to implement the proposed approach in real time.

3.3.2 Localization with Stereo Visual Odometry

For SVO data, the `libviso2` package is used to obtain raw measurements [Geiger et al., 2011b]. It is assumed that the scale of SVO is unknown, and the initial range of the scaling factor is set as [0.5, 1.5]. The estimated trajectory and localization error of our method and SVO are shown in Figure 3.2 and Table 3.1.

Qualitatively, the SVO drift increases whenever the vehicle turns sharply, as depicted in the left column of Figure 3.2. The trajectories generated from the map-assisted approach can be found in the middle column of Figure 3.2. In contrast, for the proposed approach, performance would be better for a raw VO trajectory with more turnings since the shape matching is more effective for "complicated" trajectories that indicate the vehicle's unique route.

Pure SVO localization leads to unbounded translation drift with the growth of travel distance, mainly due to the iterative nature of odometry. According to the quantitative results, the proposed approach achieved an average localization error of 6.58 m, while the error of SVO was 40.51 m. It is also observed that the proposed approach provides more robust positioning compared to pure SVO according to the lower standard deviations of the localization errors.

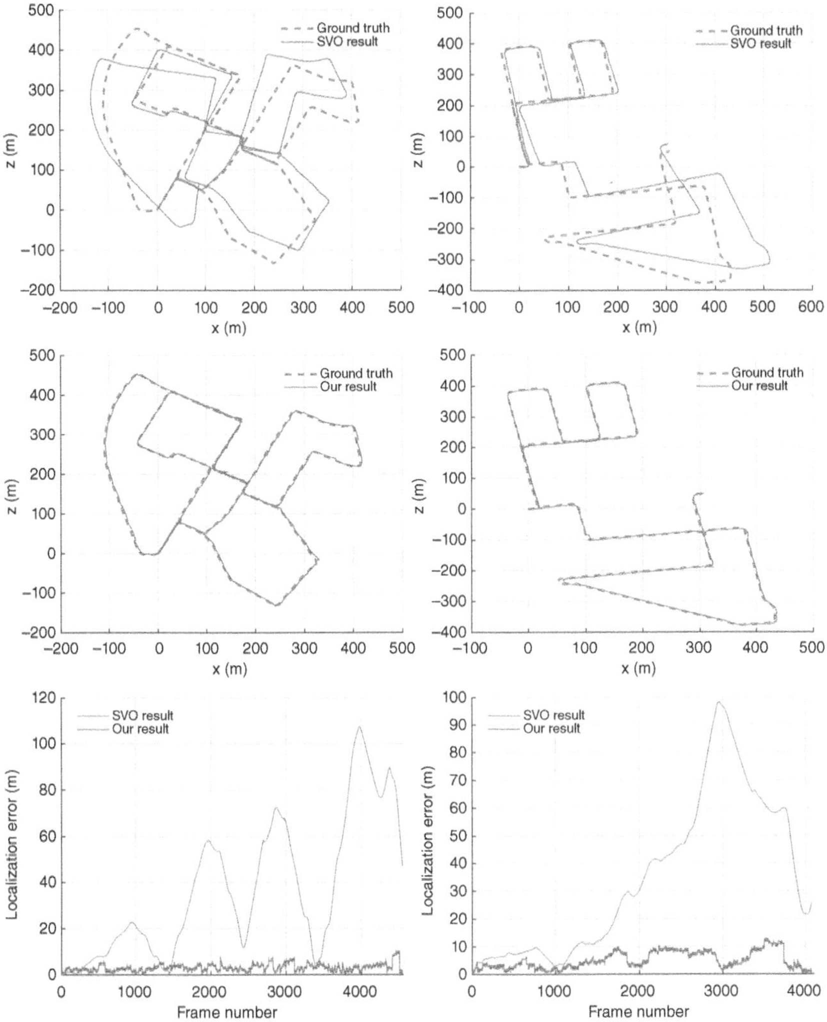

Figure 3.2 Sequences 00 and 08 from the VO benchmark of the KITTI dataset. Upper row: trajectory estimated from SVO and ground truth. Middle row: trajectory estimated from our map-assisted approach and ground truth. Lower row: localization error comparison between SVO and our method.

3.3.3 Localization with Monocular Visual Odometry

Next, we evaluate the proposed approach using MVO raw translation vectors without scale information. The ORB-SLAM is used to provide us with a benchmark in the monocular case [Mur-Artal et al., 2015a], where the vehicle's localization and mapping up to an absolute scale can be obtained simultaneously, benefiting from the loop closures. Without loop closure, standard local tracking and mapping would be performed such that ORB-SLAM acts as an MVO system. All the results from ORB-SLAM are called "MVO" in the second set of experiments. During the evaluations, we compare the performance of our method with the MVO method in conditions with and without loop closure. The trajectories estimated from the two methods are shown in Figure 3.3.

Since MVO may not provide position estimation with an absolute scale at all times, we rescale the MVO raw measurements by scaling the MVO trajectories during initialization to ground truth trajectories using shape-preserving transformation such that performance between MVO and the ground truth can be fairly compared, as shown in the top row of Figure 3.3. The right column of Figure 3.3 shows the localization errors of the two methods. As can be seen, MVO performs much better on sequence 00 than on sequence 08 because of the loop closures in sequence 00, while scale drift undermines performance on sequence 08. Despite the loop-closure optimization, the drift issue still exists in sequence 00. In contrast, the proposed approach always performs well regardless of the existence of loop closures. Most of the results are perfectly constrained near a road owing to the shape-matching and parameter-estimation scheme.

The quantitative results are listed in Table 3.2. Average errors in Brubaker et al. (2013a) are also listed for comparison. As KITTI 00 and KITTI 05 have loop closures, MVO localization errors for these sequences are 10.00 m and 7.21 m, respectively. However, MVO results on KITTI 02 (151.25 m), KITTI 08 (475.43 m), and KITTI 09 (90.81 m) become much worse. Our results are better: the average localization error is 6.54 m over the 15.89 km run.

Table 3.2 Quantitative comparison between the proposed approach and MVO localization in KITTI sequences 00, 02, 05, 08, and 09. (Distance in kilometers; mean and standard deviation of error in meters.)

		Error (proposed)			Error (MVO)			Error in Brubaker et al. (2013a)
Seq	**Dist**	**Avg**	**Std**	**Max**	**Avg**	**Std**	**Max**	**Avg**
00	3.72	4.11	2.40	13.60	10.00	6.62	26.07	15.6
02	5.06	9.53	4.36	21.90	151.25	72.64	256.72	8.1
05	2.20	6.28	5.34	22.90	7.21	7.20	23.31	5.6
08	3.21	4.43	3.38	18.29	475.43	408.21	1300.20	45.2
09	1.70	7.28	3.73	19.65	90.81	100.84	347.64	5.4
Total	15.89	6.54	–	–	160.07	–	–	16.72

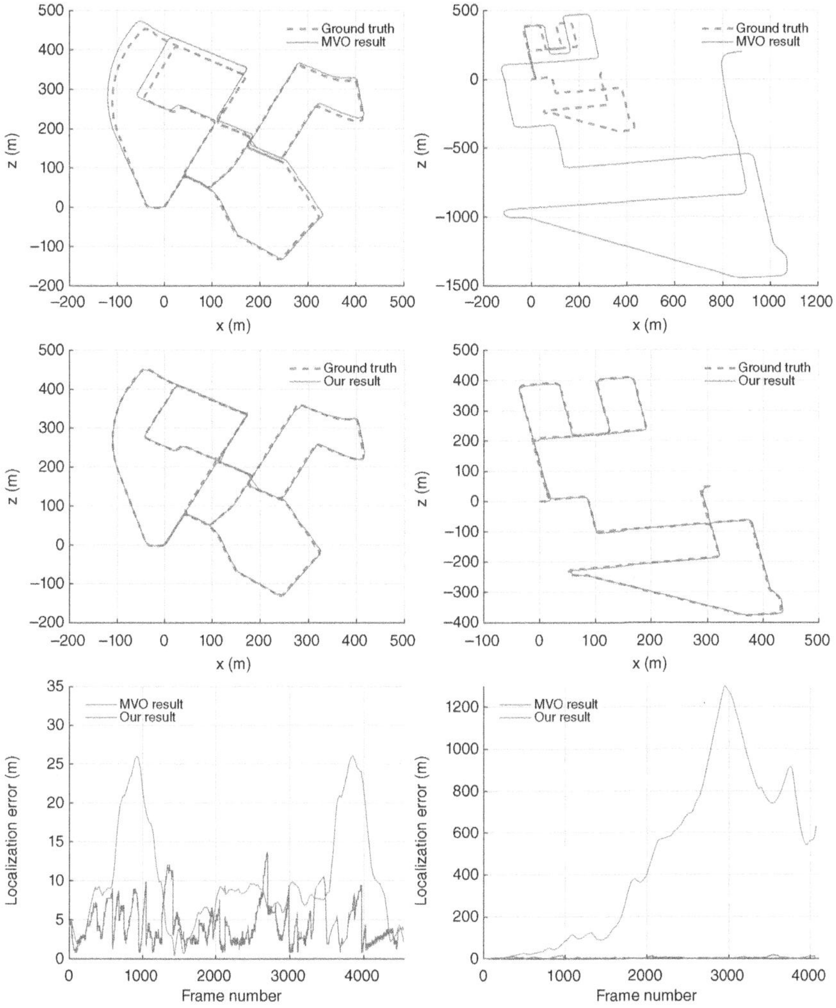

Figure 3.3 Sequences 00 and 08 from the VO benchmark of the KITTI dataset. Upper row: trajectory estimated from MVO and ground truth. Middle row: trajectory estimated from our map-assisted approach and ground truth. Lower row: localization error comparison between MVO and our method.

In short, compared to the method of Floros et al. (2013), the proposed approach is effective for both SVO and MVO; compared to the method of Brubaker et al. (2013a), the proposed approach performs worse in SVO sequences with large drifts. As the exponential drift model may not represent measurement randomness properly when the drifting error is large, it is promising to obtain less error based on a more precise drift model. Moreover, building a unified representative model with rotational error considered may also help.

Besides errors from VO and the number of particles, practical road conditions and map resolution may influence positioning performance. Error from road conditions cannot be decreased because roads are modeled with center lines, while real roads have different widths. Theoretically, error from map preparation precision can be eliminated if the resolution of the map is high enough. However, this contradicts with execution time and memory usage, as shape matching is needed for each possible trajectory to obtain particle weights.

3.3.4 Scale Estimation Results

The scale estimation results for sequences 00 and 02 are shown in Figure 3.4. Although it has not been proven that the proposed approach ensures the convergence of the estimated scale, in most cases, true scales can be recovered with road constraints. The estimated scale range is narrower when the trajectory contains more geometric information, i.e. turns. To avoid unnecessary estimation error from redundant constraints such as straight roads, scale estimation results based on straight roads are not used.

3.3.5 Weighting Function Balancing

As the weighting function in Eq. (3.19) considers both *saliency* and *consistency*, it is necessary to examine the influence of these two terms on algorithm performance. For *saliency* set $\{\mu_{\mathrm{VO}}(\mathbf{y}_i)\}$, it is obvious that

$$\min \lim_{N\to\infty} \{\mu_{\mathrm{VO}}(\mathbf{y}_i)\} = 0$$

$$\max \lim_{N\to\infty} \{\mu_{\mathrm{VO}}(\mathbf{y}_i)\} = 1$$

where N denotes the number of particles. However, this is not the case for *consistency* even after normalization, as the shape-matching distance of all particles may not cover a specific interval, especially when the estimated trajectory is off the road. This unbalanced weighting strategy may cause *consistency* to be less significant and decrease the algorithm's robustness. In practice, the following affine transform is used for *consistency* such that the distribution shape remains unchanged:

$$\mu'_{\mathrm{SM}}(\mathbf{y}_i) = \frac{\mu_{\mathrm{SM}}(\mathbf{y}_i) - \min\{\mu_{\mathrm{SM}}(\mathbf{y}_i)\}}{\max\{\mu_{\mathrm{SM}}(\mathbf{y}_i)\} - \min\{\mu_{\mathrm{SM}}(\mathbf{y}_i)\}} \tag{3.24}$$

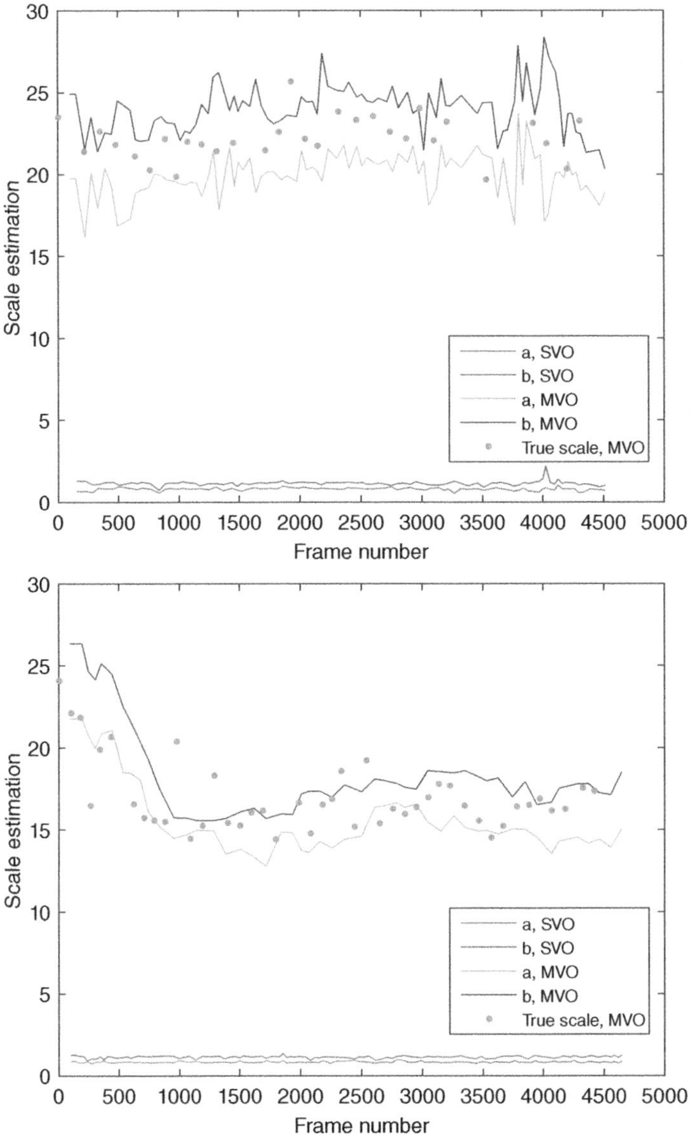

Figure 3.4 Scale estimation results for sequences 00 and 02, respectively.

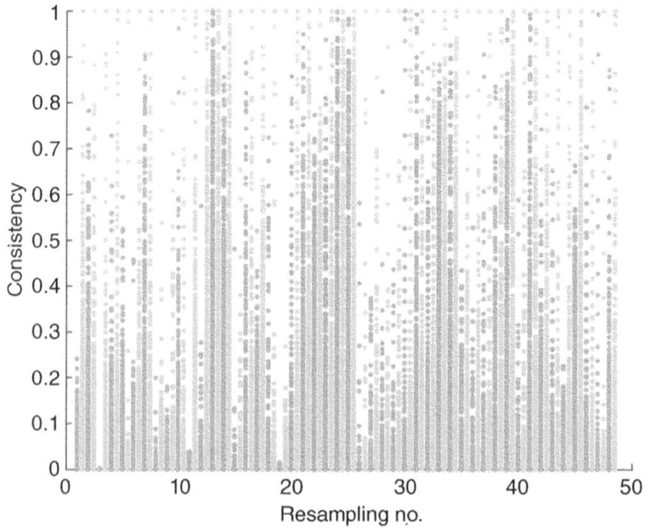

Figure 3.5 *Consistency* range comparison with (light gray) and without (dark gray) weighting function balancing.

Figure 3.5 shows the *consistency* range with and without Eq. (3.24) on MVO sequence 02. The affine transform ensures $\min \lim_{N\to\infty}\{\mu_{SM}(\mathbf{y}_i)\} = 0$ and $\max \lim_{N\to\infty}\{\mu_{SM}(\mathbf{y}_i)\} = 1$. It is noted that the weighting function after balancing avoids the possible significant degradation problem of *consistency*.

3.4 Experimental Validation on GGD

To evaluate the performance of the proposed approach in real conditions, experiments were conducted on the KITTI VO dataset as well as our self-collected dataset. These two datasets are captured by driving a vehicle around urban and suburban environments at normal driving speed.

Even though the original intention of this work was to tackle measurement uncertainties of MVO, which include drift and scale ambiguity, localization with both stereo and monocular VO were conducted, and the results are shown to demonstrate the performance of our proposed approach. Some typical parameters are set as follows: initialization criterion $d_{\min} = 300$ m; $\alpha_{\min} = \pi/4$; weight $\alpha = 0.5$ for orientation distance in shape matching; degrees of freedom $k = 3$ for Chi-squared distribution; 800 drops sampled in the proposed framework. All the experiments were conducted on a mobile workstation with an i7-4710MQ processor.

3.4.1 Experiments on KITTI

The KITTI VO dataset contains 22 stereo sequences, and the ground truth has been released for the first 11. These sequences are made with various lengths and trajectory shapes; thus, they are suitable for localization performance evaluation. Some of them were used in this experiment.

Localization with Stereo Visual Odometry In the first round of experiments, the proposed approach is evaluated with SVO dead-reckoning results. We choose the widely used `libviso2` package [Geiger et al., 2011b] as an SVO model. During the experiment, it is assumed that the scale of SVO is unknown and the initial scaling factor satisfies $\mathbf{S} \sim N(1, 0.5)$. In that sense, SVO can be regarded as a special MVO.

Some representative trajectories estimated from SVO and our road-constrained approach, as well as the localization error comparison curve, can be found in Figure 3.6. As can be seen, the SVO drift increases every time the robot turns sharply. And due to the accumulative drift, pure SVO localization becomes

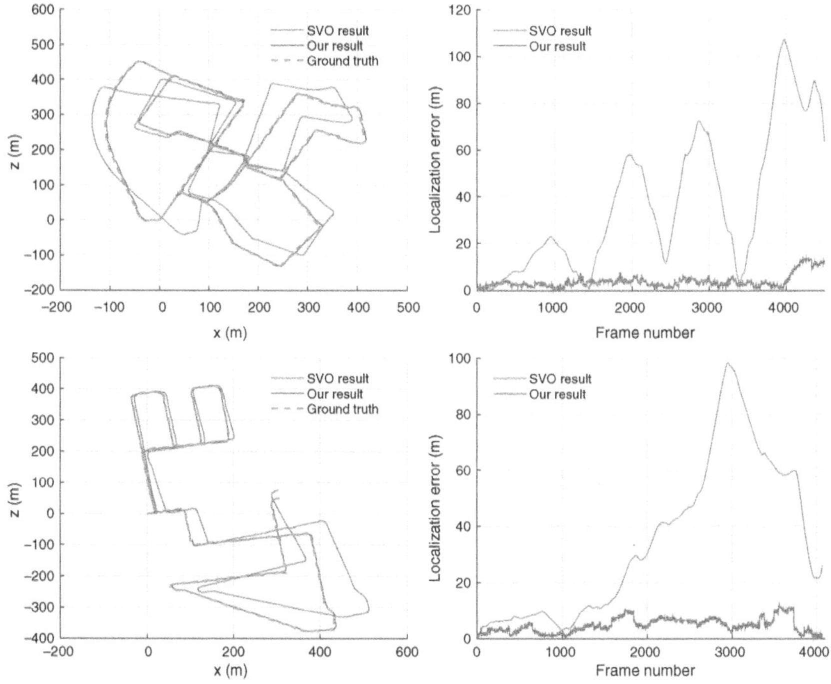

Figure 3.6 Sequences 00 and 08 from visual odometry benchmark of the **KITTI** dataset. First column: trajectories estimated from stereo visual odometry, our road-constrained algorithm, and ground truth. Second column: localization error comparison between SVO and our method.

increasingly unreliable with the growth of travel distance. In contrast, for the proposed approach, the more complicated the trajectory is, the better the localization results will be. It is not difficult to understand this unique phenomenon, as every time the robot changes its direction of motion, more information is added to the robot's trajectory, and the shape-matching algorithm narrows down the possible region of the robot.

Quantitative results from our method, pure SVO, and the state-of-the-art map-aided approach **Lost** [Brubaker et al., 2013b] are listed in Table 3.3. Noticed that not all 11 sequences of the KITTI dataset are listed in the table. That is because some sequences do not satisfy the algorithm initialization criterion, as discussed in Yang et al. (2017b). According to the results, both our method and **Lost** perform much better than pure SVO, as expected. The proposed approach reduced the average localization error drastically, from 160.07 m to 6.59 m. In addition, the proposed approach provides more robust positioning than pure SVO, as the standard deviations for all sequences are lower. Furthermore, the success of our SVO-based localization proves the effectiveness of the proposed framework, especially the effectiveness of our parameter estimation scheme. Finally, even though **Lost** has slightly better performance than the proposed approach, the advantages of our approach will be demonstrated in the monocular case.

Localization with Monocular Visual Odometry In the second round of experiments, we evaluate the proposed approach using the raw translation vectors of MVO without scale information. A state-of-the-art monocular SLAM system (ORB-SLAM) [Mur-Artal et al., 2015b] is used to provide us with monocular motion prediction.

Table 3.3 Quantitative comparison in localization errors between the proposed approach, pure SVO, and Brubaker et al. (2013) in KITTI sequences 00, 02, 05, 08, and 09 (distance in kilometers; mean and standard deviation of error in meters).

		Proposed			SVO			[Brubaker et al., 2013]
Seq	**Dist**	**Avg**	**Std**	**Max**	**Avg**	**Std**	**Max**	**Avg**
00	3.72	3.76	2.80	14.01	37.11	29.63	107.23	2.1
02	5.06	11.32	7.51	25.60	66.08	47.37	172.71	4.1
05	2.20	4.02	2.04	8.70	14.40	13.30	45.60	2.6
08	3.21	4.67	2.57	11.96	34.70	28.12	98.20	2.4
09	1.70	5.72	2.67	11.57	16.63	34.91	9.72	4.2
Total	15.89	6.59	–	–	40.51	–	–	3.19

To validate the effectiveness of our proposed method, we compare the performance of our method with the MVO method both with loop closure and without loop closure. Since the absolute scale is unknown, to compare monocular motion estimation with the ground truth, we rescale the MVO raw measurements by aligning the odometry trajectories during initialization with the ground truth using a similarity transformation, as shown in the first row of Figure 3.7. As can be

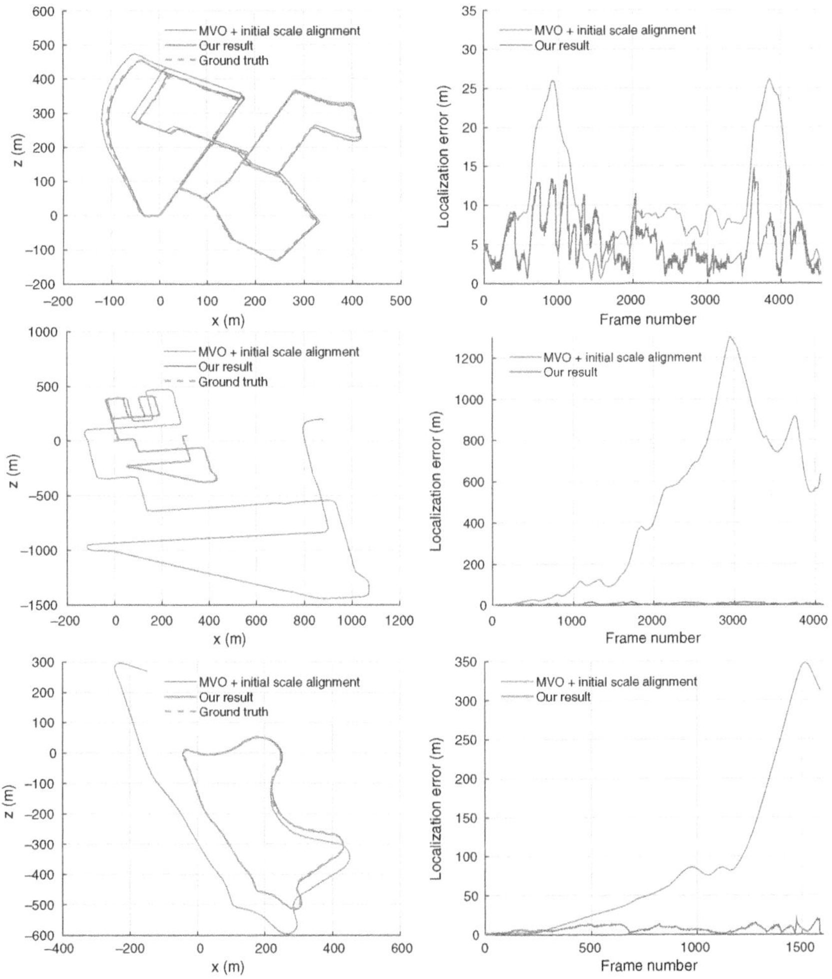

Figure 3.7 Sequences 00, 08, and 09 from the visual odometry benchmark of the **KITTI** dataset. First column: trajectories estimated from monocular visual odometry aligned with the initial scale, our road-constrained algorithm, and ground truth. Second column: localization error comparison between MVO and our method.

seen, MVO performs much better on sequence 00 than on sequences 08 and 09 because loop closures exist in route 00 but not in the other two sequences (actually, sequence 09 has loop closures at the very end, but ORB-SLAM fails to detect them). The 7DoF drifts, especially the scale drift, destroy the motion estimation for sequences 08 and 09. Despite the loop-closure optimization, the drift issue still exists in sequence 00. In contrast, our road-constrained method always performs well regardless of loop closures. Most of the results are perfectly constrained on or near a road, thanks to our shape-matching and parameter estimation scheme. The second row of Figure 3.7 shows the error curves of the two methods. Based on the localization error comparison, it can be concluded that the proposed framework is valid for monocular localization.

The quantitative results are also listed in Table 3.4. As KITTI sequences 00 and 05 have loop closures, MVO results for these sequences are still acceptable. However, MVO results from sequences 08 and 09 are not as optimistic as the first two sequences. At the same time, the results from our algorithm are much better, as the average localization error is 5.62 m compared to 160.07 m over the 10.83 km run. Moreover, the proposed approach outperforms **Lost** in this monocular case. As can be seen, our MVO-based localization has roughly the same performance as our SVO-based approach, while **Lost** has greater inconsistency. This further proves the effectiveness of the proposed for both SVO and MVO.

In order to demonstrate the advantage of using the cloud model with respect to models that consider the two ambiguities separately, comparison experiments are done. Concretely, scale ambiguity is modeled, while MVO measurement randomness is not considered. In this scenario, the scaled measurement follows $GGC(E\mathbf{S}, \Sigma_S, E\mathbf{X}, \mathbf{0})$. The result is listed as the next-to-last column of Table 3.4. As can be seen, the positioning error of this model is larger than the complete model. It is not hard to understand this result: since measurement randomness is not properly modeled here, translational and rotational drift issues still exist. An experiment evaluating another scenario is also conducted. This time, measurement randomness is modeled, while scale ambiguity is not considered. The scaled measurement follows $GGC(E\mathbf{S}, \mathbf{0}, E\mathbf{X}, \Sigma_X)$ in this case. The result is listed in the last column of Table 3.4. Unsurprisingly, this model performs much worse than the complete model due to the unhandled scale-drift issue.

Scale-Drift Correction Figure 3.8 shows the absolute scale estimated from the proposed method. Light gray and solid dark gray lines are the estimated scales of stereo and monocular odometry, respectively. The "true" scale at a certain distance - indicated with a dark gray cross - is calculated from the ratio between ground truth and the raw odometry measurement. As can be seen, stereo and

Table 3.4 Quantitative results of errors between the proposed approach, pure MVO aligned with initial scales, Brubaker et al. (2013), model only considering scale ambiguity, and model only considering odometry measurement randomness. Sequences used are KITTI 00, 05, 08, and 09 (distance in kilometers; mean and standard deviation of error in meters).

		Proposed			MVO			[Brubaker et al., 2013b]	Only Ambiguity	Only Randomness
Seq	**Dist**	**Avg**	**Std**	**Max**	**Avg**	**Std**	**Max**	**Avg**	**Avg**	**Avg**
00	3.72	5.43	3.15	14.70	10.00	6.62	26.07	15.6	6.01	9.2
05	2.20	6.02	4.82	18.70	7.21	7.20	23.31	5.6	8.13	7.0
08	3.21	5.05	3.23	14.67	475.43	408.21	1300.20	45.2	66.9	462.3
09	1.70	6.63	2.93	16.45	90.81	100.84	347.64	5.4	63.4	85.3
Total	10.83	5.62	–	–	160.07	–	–	16.72	33.5	154.9

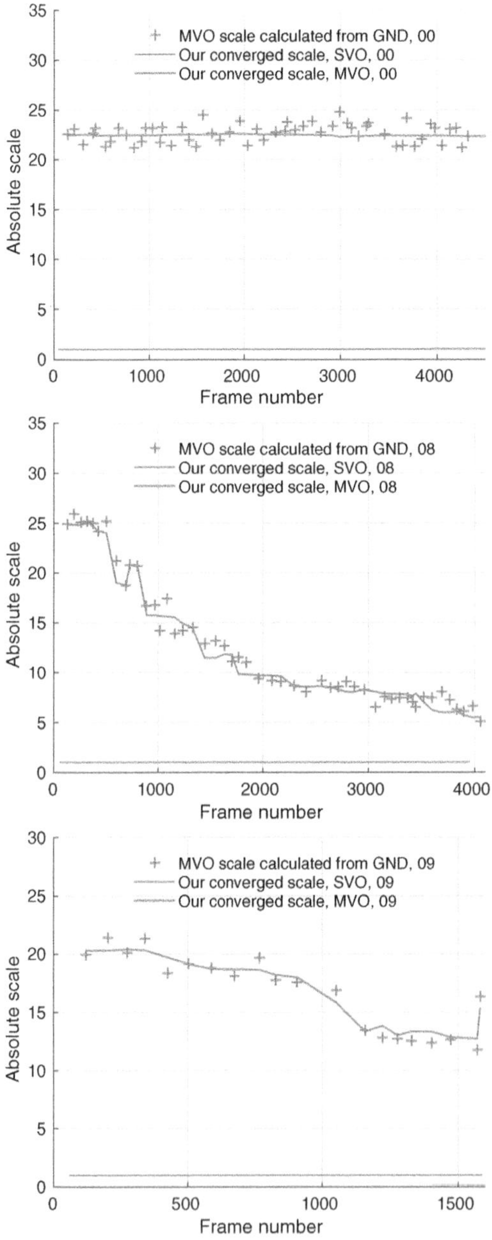

Figure 3.8 The absolute scales of sequences 00, 08, and 09. The light gray solid lines represent corrected scales for stereo visual odometry. The dark gray solid lines represent corrected scales for monocular visual odometry. Dark gray crosses show the scales calculated from ground truth.

monocular odometry perform differently, and different sequences also perform diversely. Most of the stereo sequences converge successfully to the ideal scale, which is precisely what we need to verify the feasibility of the proposed parameter updating scheme. For the monocular case, sequence 00 has a nearly constant scale, while sequences 08 and 09 have greatly drifted scales. Although the converged scales are not exactly the same as the scales calculated from ground truth, our results fit the changing tendency well. Please note the huge variation in the scales of sequences 08 and 09, which further verifies the effectiveness of our parameter estimation.

3.4.2 Experiments on the Self-Collected Dataset

In this section, experiments conducted on our self-collected dataset are described. Our evaluation platform is a Honda CR-V. A stereo camera set is mounted on the roof of the vehicle and oriented forward. It is configured to acquire stereo frames at 50 Hz with a resolution of 1280×1024. The vehicle is also equipped with GPS to provide us with position ground truth. Our datasets are captured by driving around the campus of Nanyang Technological University.

Similar to the first experiment on the KITTI dataset, both stereo and monocular cases are considered here, and the same VO packages are utilized. Trajectories estimated from our method, as well as trajectories estimated from pure SVO and MVO systems are shown in Figure 3.9. Unsurprisingly, poor results are obtained from pure odometry systems. Moreover, SVO performs worse than MVO since no loop-closing technique is used. In the meantime, our proposed method performs very well, as always. The rightmost column of Figure 3.9 shows the absolute scale estimated from our method. Obviously, the stereo case can still converge to one correctly. But for the monocular case, two sudden changes of scale occur around the 500th and 2200th frames when the vehicle makes turns.

Experiments analyzing the importance of parameter setting were also conducted. The two curves in Figure 3.10 demonstrate the influence of the given parameters (ES, σ_s) on convergence performance. As can be seen in Figure 3.10(a), the estimated scale can converge to the truth when the given scale's expectation lies in the rough range $[ES_{truth} - \sigma_s, ES_{truth} + \sigma_s]$. It will always be possible to converge to the true scale as long as the given standard deviation is large enough. Nevertheless, as shown in Figure 3.10(b), the convergence time used to converge to the true position increases exponentially over the given standard deviation, which verifies the significance of narrowing the scale ambiguity when resampling drops.

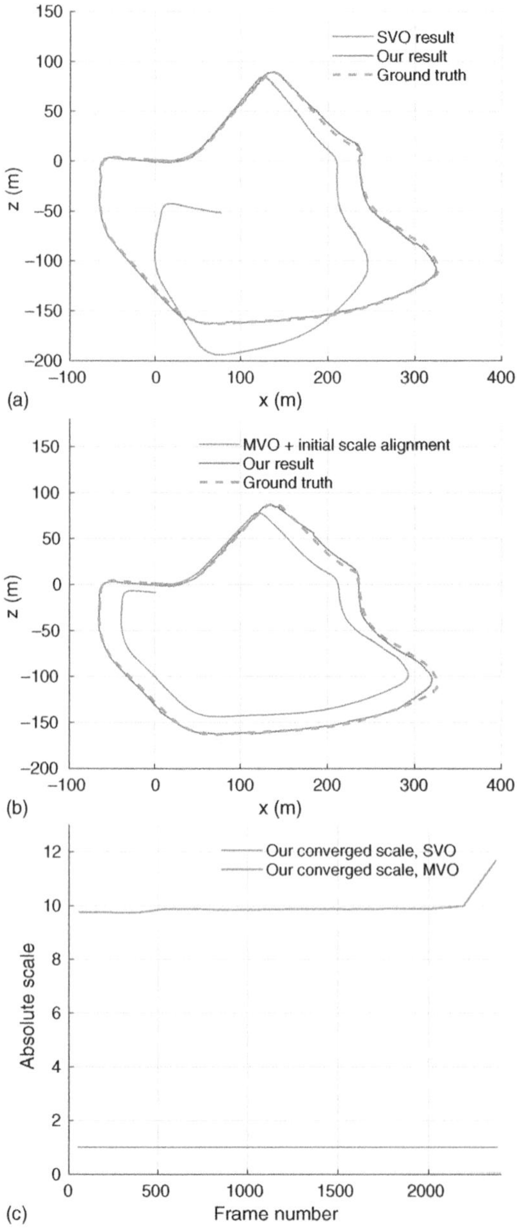

Figure 3.9 Our self-collected dataset. (a) Trajectories estimated from stereo visual odometry, our road-constrained algorithm, and ground truth. (b) Trajectories estimated from monocular visual odometry aligned with initial scale, our road-constrained algorithm, and ground truth. (c) Scales estimated from our method.

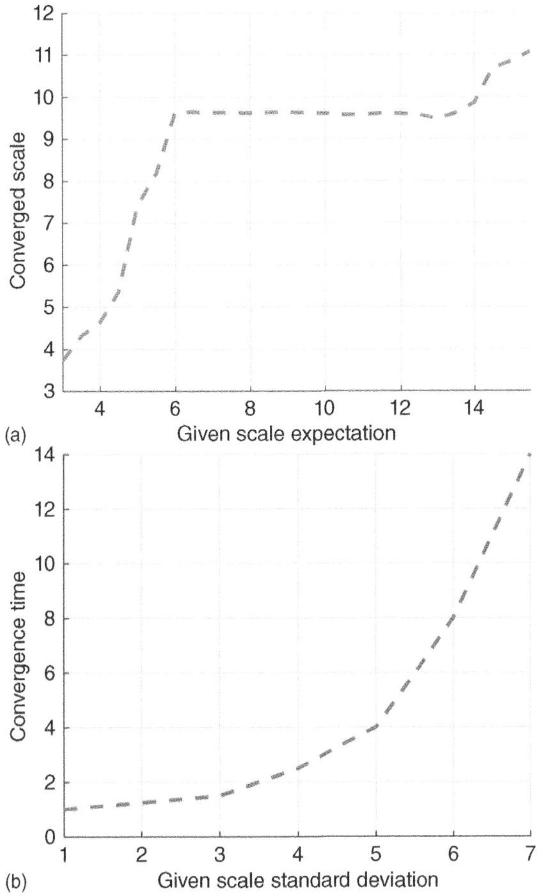

Figure 3.10 The influence of the given parameters. (a) The influence of given scale exception *ES* on the final converged scale; the true scale is 9.7, $\sigma_s = 3.5$. (b) The influence of given scale standard deviation σ_s on the convergence time; the smallest time consumption is considered one unit.

3.5 Conclusion

In this chapter, a map-assisted localization approach has been presented that utilizes the dead-reckoning trajectory from VO and a map to obtain position estimation. Two new probability distributions, the uniform Gaussian distribution and the Gaussian-Gaussian distribution, have been proposed to describe the measurement uncertainties in VO, and a framework has been designed to incorporate the motion *saliency* concerning VO and the measurement *consistency* to the road shape. Experiments have shown that with road constraints, the translational localization error has been significantly reduced compared to an advanced MVO method.

4

GPS/Odometry/Map Fusion for Vehicle Positioning Using Potential Functions

Data fusion is a popular approach to tackling vehicle localization problems. For vehicles in complex environments, a single information source is still far from enough for positioning accuracy and robustness. For example, as an absolute positioning sensor, a global positioning system (GPS) performs well in large-scale localization but suffers from temporary signal loss and small-scale inaccuracy [Huang and Pi, 2011, Kong, 2011]. Odometry and inertial sensors are based on dead reckoning or relative positioning, whose accumulative error makes the estimated location drift over time [Borenstein and Feng, 1996, Barshan and Durrant-Whyte, 1995]. Vision-based approaches generally match landmarks in the current frame with database maps [Se et al., 2005, Ramisa et al., 2009], while the requirement of a pre-established database limits its application. Digital maps that provide constraints for the vehicle on the road are increasingly used as an additional information source, especially in urban areas [El Najjar and Bonnifait, 2005, Alonso et al., 2012, Jiang et al., 2017], and map matching is usually a necessary procedure to align other measurements with roads.

In previous work, most research is on the lower levels of fusion [Smith and Singh, 2006], and several fusion methods are proposed for vehicle positioning. So far, the Kalman filter and its varieties (e.g. extended Kalman filter, information filter, sigma point Kalman filter, and multiple model algorithms) are still the most popular fusion methods [Noureldin et al., 2009, Wei et al., 2013, Crassidis, 2006, Jo et al., 2012]. The Kalman filter-related approaches need information about the system model and process noise, which sometimes is unavailable. When information conflicts, it is hard to detect the conflict and provide reasonable positioning results immediately. Moreover, the Kalman filter-related approaches are unsuitable for fusing data with distinct properties. For instance, it is difficult to ruse a road model (generally represented by a set of points on digital maps) and a position measurement (represented by a point with a timestamp) using a

Multimodal Perception and Secure State Estimation for Robotic Mobility Platforms, First Edition.
Xinghua Liu, Rui Jiang, Badong Chen, and Shuzhi Sam Ge.

Kalman filter, as the road model only describes the vehicle's position over a period of time and cannot be substituted into the filtering scheme as a measurement at a certain time.

In this chapter, we introduce a fusion approach by introducing the concept of potential functions to form potential wells and potential trenches. Under this fusion frame, the data fusion problem can be converted to a minimum searching problem, where each information source produces a corresponding potential field according to data properties, and the spatial point with the lowest potential is estimated as a fused position. The main highlights of the chapter are as follows:

- An insightful and straightforward data fusion frame based on potential functions is proposed and implemented in the vehicle positioning problem. Visual odometry (VO), GPS, and 2D digital road maps are integrated inherently without map-matching algorithms.
- The fusion scheme provides realistic results when the measurements do not completely satisfy the constraints. Thus the robustness of the system is enhanced.
- The proposed scheme does not rely on a kinematic model and can be easily generalized to other data fusion problems.

The chapter proceeds as follows. In Section 4.1, the concepts of potential functions, potential wells, and potential trenches are introduced, followed by the elaboration of potential field creation according to information properties. Section 4.2 details the potential-function-based fusion approach for vehicle positioning; the experimental results are presented in Section 4.3. Finally, conclusions are offered in Section 4.4.

4.1 Potential Wells and Potential Trenches

The concept of an artificial potential field was first proposed in Khatib (1986) to deal with obstacle avoidance for manipulators and mobile robots; and potential trenches were introduced in Ge and Fua (2005) in multi-robot formations, where goals create attractive fields and obstacles create repulsive fields. The sum of attractive fields and repulsive fields makes the robot move in configuration space. In this chapter, the sources of potential fields are data, and the fusion results are searched for in the data space. Before proceeding further, we define several concepts.

Definition 4.1 ***(Data space)*** An n-dimensional data space in the fusion frame D^n is the n-dimensional vector space of all possible data and fusion results under the frame.

We only consider fusion problems on a single level [Hall and Llinas, 2001]; thus, data and fusion results should be compatible in one data space. Data and fusion results in a data space can be represented as geometric shapes, such as points, lines, surfaces, or hyper-surfaces. For example, in a 2D data space, data measured by a position sensor can be represented as a point (x, y), and an on-the-road constraint can be expressed as a curve $c(x, y) = 0$.

Definition 4.2 ***(Potential function [Choset, 2005])*** A potential function in n-dimensional data space is a differentiable real-valued function $U : \mathbb{R}^n \to \mathbb{R}$. Given a vector field $\mathbf{F}$, the potential U is defined such that

$$\mathbf{F} = -\nabla U \tag{4.1}$$

Both potential function U and vector field $\mathbf{F}$ can be used to describe a data space. Considering the convenience of scalar operations, we mainly design U according to data properties. Two types of potential functions are defined in this chapter: potential wells and potential trenches.

Definition 4.3 ***(Potential well functions)*** Given a point $\mathbf{q}^{\mathrm{s}} = (q_1^{\mathrm{s}}, q_2^{\mathrm{s}}, \dots, q_n^{\mathrm{s}})^T$ in $\mathcal{D}^n$, the potential well function is defined as $U_{\mathrm{w}}(\mathbf{q}) = \alpha_{\mathrm{w}} f_{\mathrm{w}}(\mathbf{d}_{\mathrm{w}})$, where $\alpha_{\mathrm{w}} > 0$ is a user-defined parameter, $f_{\mathrm{w}}(\cdot)$ is the potential construction function, and $\mathbf{d}_{\mathrm{w}} = \mathbf{q} - \mathbf{q}^{\mathrm{s}}$ is the distance vector between $\mathbf{q}^{\mathrm{s}}$ and $\mathbf{q}$. The point $\mathbf{q}^{\mathrm{s}}$ is called the source of the potential well.

Definition 4.4 ***(Distance between a shape and a point)*** Given a shape $c^{\mathrm{s}}(\mathbf{q}) = 0 \in \mathcal{D}^n$ and a point $\mathbf{q} = (q_1, q_2, \cdots, q_n)^T$, the distance between the shape $c^{\mathrm{s}}(\mathbf{q}) = 0$ and the point $\mathbf{q}$ is defined as the radius of the hyper-sphere centered at $\mathbf{q}$ and tangent to $c^{\mathrm{s}}(\mathbf{q}) = 0$.

Definition 4.5 ***(Potential trench functions)*** Given a shape $c^{\mathrm{s}}(\mathbf{q}) = 0 \in \mathcal{D}^n$, the potential trench function is defined as $U_{\mathrm{t}}(\mathbf{q}) = \alpha_{\mathrm{t}} f_{\mathrm{t}}(d_{\mathrm{t}})$, where $\alpha_{\mathrm{t}} > 0$ is a user-defined parameter, $f_{\mathrm{t}}(\cdot)$ is the potential construction function, and d_{t} is the distance between the shape $c^{\mathrm{s}}(\mathbf{q}) = 0$ and the point $\mathbf{q}$. The shape $c^{\mathrm{s}}(\mathbf{q}) = 0$ is called the source of the potential trench.

To visually illustrate a potential well and a potential trench, two examples are created in Figure 4.1.

4.1.1 Potential Function Creation

Based on the previous definitions, in order to create a potential well $U_{\mathrm{w}}(\mathbf{q}) = \alpha_{\mathrm{w}} f_{\mathrm{w}}(\mathbf{d}_{\mathrm{w}})$ and a potential trench $U_{\mathrm{t}}(\mathbf{q}) = \alpha_{\mathrm{t}} f_{\mathrm{t}}(d_{\mathrm{t}})$ in data space, it is necessary to

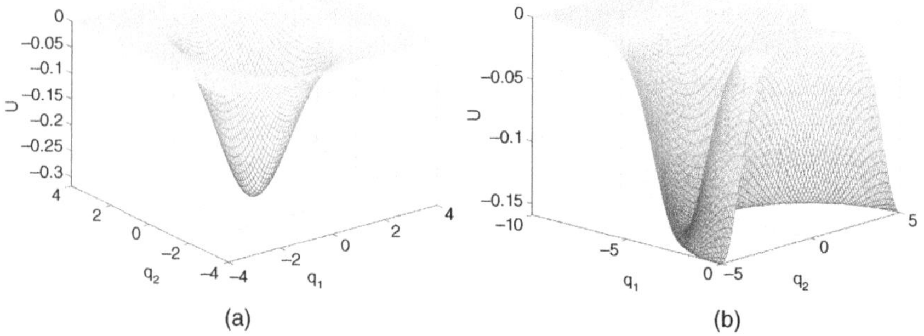

Figure 4.1 Examples of a potential well and a potential trench in 2D data space. (a) Potential well with $\alpha_w = 1$, construction function $f(\mathbf{d}_w) = -(2\pi)^{-1}|\mathbf{D}|^{-\frac{1}{2}} \exp\left(-\frac{1}{2}\mathbf{d}_w^T\mathbf{D}^{-1}\mathbf{d}_w\right)$, and source $\mathbf{q}^s = (0, 0)$ and $\mathbf{D} = \text{diag}(1, 2)$. (b) Potential trench with $\alpha_w = 1$, construction function $f(d_t) = -\sigma^{-1}(2\pi)^{-\frac{1}{2}} \exp\left(-\frac{d_t^2}{2\sigma^2}\right)$, and source $c^s(q_1, q_2) = q_1^2 + q_2^2 - 25 = 0$ and $\sigma = 1$.

determine the parameters α_w, α_t, sources $\mathbf{q}^s$, $c^s(\mathbf{q}) = 0$, and potential construction functions $f_w(\cdot)$, $f_t(\cdot)$. The potential construction functions map the distance between a source and a point in data space to a real number. According to Definitions 4.2, 4.3, and 4.5, given potential functions $f_w(\cdot)$ and $f_t(\cdot)$, $U_w(\mathbf{q})$ and $U_t(\mathbf{q})$ can be obtained self-consistently. Without loss of generality, the default potential field in data space is initialized to zero, and the point with minimum potential is regarded as the fusion result. Thus, $f_w(\cdot) \leq 0$ and $f_t(\cdot) \leq 0$ should be ensured, and the potentials with $d_t = 0$ or $\mathbf{d}_w = \mathbf{0}$ are required to be the minimum for the corresponding potential functions. To summarize, any form of f can be created if f satisfies the following: (i) f is a potential function and (ii) $f \leq 0$; 3) f reaches the minimum at 0.

Due to the distinct properties between potential wells and trenches, we will discuss how to create f with regard to measurements and constraints separately.

Potential Wells as Measurements A measurement can be represented as a point in data space, where the source of a potential well is also a point. It is natural to create potential wells representing measurements from different sensors. There are basically two ways to create potential wells: (i) creating multiple potential wells at a time instant, one for each information source; and (ii) creating an individual potential well for all measurement data. In this chapter, the second approach is preferred, as the first method leads to a mixture model that does not conform to reality, and an individual potential contributes to local minimum avoidance. The following three assumptions are made before we proceed:

Assumption (*Synchronous fusion*) The data from different information sources is synchronized. □

Assumption The data from information source i at the same time follows a Gaussian distribution with mean $\mathbf{q}_i^s$ and covariance matrix $\mathbf{D}_i$. □

Assumption The fused data $\mathbf{y}$ can be represented as the linear combination of different independent data sources: $\mathbf{y} = \sum_{i=1}^{m} \mathbf{W}_i \mathbf{y}_i$, where $\mathbf{W}_i$ denotes weights satisfying $\sum_{i=1}^{m} \mathbf{W}_i = \mathbf{I}$, $\mathbf{y}_i$ $(i = 1, \cdots, m)$ are measurement data from m independent information sources, and $\mathbf{I}$ is the identity matrix. □

According to the assumptions, it follows that

$$\mathbf{y}_i \sim N(\mathbf{q}_i^s, \mathbf{D}_i) \tag{4.2}$$

where $\mathbf{q}_i^s$ is the unknown true value, and $\mathbf{D}_i$ is the covariance matrix for each information source. The matrix $\mathbf{D}_i$ can be estimated from sensor properties or observation data. It has been proven that

$$\mathbf{y} \sim N\left(\sum_{i=1}^{m} \mathbf{W}_i \mathbf{q}_i^s, \sum_{i=1}^{m} \mathbf{W}_i^2 \mathbf{D}_i\right) \sim N(\mathbf{q}^s, \mathbf{D}) \tag{4.3}$$

As we already know $\mathbf{D}_i$, the problem turns to estimating $\mathbf{q}^s$ and $\mathbf{W}_i$ in order to obtain fused data.

Proposition 4.1 *Given data $\mathbf{y}_i$ and covariance matrix $\mathbf{D}_i (i = 1, \cdots, m)$ from m independent information sources, the fused data $\mathbf{y}$ can be obtained from*

$$\mathbf{y} = \hat{\mathbf{q}}^s = \sum_{i=1}^{m} \mathbf{W}_i \mathbf{y}_i \tag{4.4}$$

with weights

$$\mathbf{W}_i = \left(\sum_{i=1}^{m} \mathbf{D}_i^{-1}\right)^{-1} \mathbf{D}_i^{-1} \tag{4.5}$$

Proof: According to the assumptions, the fused data follows a Gaussian distribution. By applying the maximum likelihood method, we may estimate $\mathbf{q}^s$ from the maximizing likelihood function

$$L(\mathbf{q}^{\mathrm{s}}|\mathbf{y}_1,\cdots,\mathbf{y}_m)=\prod_{i=1}^{m}p(\mathbf{y}_i|\mathbf{q}^{\mathrm{s}},\mathbf{D}_i) \tag{4.6}$$

where $p(\mathbf{y}_i|\mathbf{q}^{\mathrm{s}},\mathbf{D}_i)=(2\pi)^{-\frac{n}{2}}|\mathbf{D}_i|^{-\frac{1}{2}}\exp\left(-\frac{1}{2}(\mathbf{y}_i-\mathbf{q}^{\mathrm{s}})^T\mathbf{D}_i^{-1}(\mathbf{y}_i-\mathbf{q}^{\mathrm{s}})\right)$.
By substituting $p(\mathbf{y}_i|\mathbf{q}^{\mathrm{s}},\mathbf{D}_i)$ into Eq. (4.6), the likelihood function is written as

$$L(\mathbf{q}^{\mathrm{s}}|\mathbf{y}_1,\cdots,\mathbf{y}_m)=(2\pi)^{-\frac{mn}{2}}\prod_{i=1}^{m}|\mathbf{D}_i|^{-\frac{1}{2}}\exp\left(-\frac{1}{2}\sum_{i=1}^{m}(\mathbf{y}_i-\mathbf{q}^{\mathrm{s}})^T\mathbf{D}_i^{-1}(\mathbf{y}_i-\mathbf{q}^{\mathrm{s}})\right) \tag{4.7}$$

The log-likelihood function is

$$\ln(L)=\ln\left((2\pi)^{-\frac{mn}{2}}\right)+\sum_{i=1}^{m}\left(\ln\left(|\mathbf{D}_i|^{-\frac{1}{2}}\right)-\frac{1}{2}(\mathbf{y}_i-\mathbf{q}^{\mathrm{s}})^T\mathbf{D}_i^{-1}(\mathbf{y}_i-\mathbf{q}^{\mathrm{s}})\right) \tag{4.8}$$

Taking its derivative with regard to $\mathbf{q}^{\mathrm{s}}$ and setting it to zero, we have

$$\frac{\mathrm{d}\ln(L)}{\mathrm{d}\mathbf{q}^{\mathrm{s}}}=-\frac{1}{2}\frac{\mathrm{d}}{\mathrm{d}\mathbf{q}^{\mathrm{s}}}\left(\sum_{i=1}^{m}(\mathbf{y}_i-\mathbf{q}^{\mathrm{s}})^T\mathbf{D}_i^{-1}(\mathbf{y}_i-\mathbf{q}^{\mathrm{s}})\right) \tag{4.9}$$

$$=\sum_{i=1}^{m}\mathbf{D}_i^{-1}(\mathbf{y}_i-\mathbf{q}^{\mathrm{s}}) \tag{4.10}$$

$$=\mathbf{0} \tag{4.11}$$

$$\Rightarrow\hat{\mathbf{q}}^{\mathrm{s}}=\left(\sum_{i=1}^{m}\mathbf{D}_i^{-1}\right)^{-1}\sum_{i=1}^{m}\mathbf{D}_i^{-1}\mathbf{y}_i \tag{4.12}$$

□

The result is similar to a weighted average calculation. From the result we may find that i) For a specific measurement with a small covariance, the weight will be large, indicating that the measurement is more trusted. ii) For m information sources with the same covariance matrix, the weight will be equal. The fusion strategy degenerates to the simple average method.

Since we have obtained the fused measurement, we proceed to create a potential well based on fused data. We define the negative Gaussian potential well in $\mathcal{D}^n$ as

$$U_{\mathrm{w}}(\mathbf{q})=\alpha_{\mathrm{w}}f_{\mathrm{w}}(\mathbf{d}_{\mathrm{w}}) \tag{4.13}$$

$$f_{\mathrm{w}}(\mathbf{d}_{\mathrm{w}})=-\frac{1}{(2\pi)^{\frac{n}{2}}|\mathbf{D}|^{\frac{1}{2}}}\exp\left(-\frac{1}{2}\mathbf{d}_{\mathrm{w}}^T\mathbf{D}^{-1}\mathbf{d}_{\mathrm{w}}\right) \tag{4.14}$$

where $\mathbf{d}_{\mathrm{w}}=\mathbf{q}^{\mathrm{s}}-\mathbf{q}$. The covariance matrix $\mathbf{D}$ can be calculated from Eq. (4.3), and the source $\mathbf{q}^{\mathrm{s}}$ can be estimated according to Eqs. (4.4) and (4.5).

Potential Trenches as Constraints The potential trench can be created to reflect constraints in data space. We consider a single constraint first. Suppose we have the constraint $c^S(\mathbf{q}) = 0$ in $\mathcal{D}^3$, referring to Definition 4.4; the distance between $c^S(\mathbf{q}) = 0$ and $\mathbf{q}$ can be solved from the following general equation

$$d_t = \min \sqrt{(\mathbf{q}^s - \mathbf{q})^T (\mathbf{q}^s - \mathbf{q})} \quad \text{subject to } c^S(\mathbf{q}^S) = 0 \tag{4.15}$$

As the constraint is another kind of information source, which may not be absolutely trusted, we define the negative Gaussian potential trench in $\mathcal{D}^n$ as

$$U_t(\mathbf{q}) = \alpha_t f_t(d_t) \tag{4.16}$$

$$f_t(d_t) = -\frac{1}{\sigma\sqrt{2\pi}} \exp\left(-\frac{d_t^2}{2\sigma^2}\right) \tag{4.17}$$

where the confidence coefficient σ directly influences how much we trust the constraint.

Now let us consider multiple constraints. Similar to multiple data sources in potential well creation, there are two ways to deal with multiple constraints: (i) creating multiple potential trenches; and (ii) creating a single trench for the current constraint, meanwhile switching the constraint based on specific conditions varying with time. For vehicle localization, the second approach is preferred, as the vehicle is always constrained on one specific piece of road at a given moment. Also, more potential functions may decrease real-time performance and increase the chance of generating local minima. The switching strategy will be introduced in Section 4.2.3.

4.1.2 Minimum Searching

The potential field in data space may be written as the sum of all potential functions

$$U(\mathbf{q}) = U_w(\mathbf{q}) + U_t(\mathbf{q}) \tag{4.18}$$

with user-defined parameters α_w, $\alpha_t > 0$. The points with local minimum potentials in data space are regarded as the candidates of the fusion result.

The most straightforward method to find local minimum potentials is to take derivatives of $U(\mathbf{q})$ with respect to $\mathbf{q}$ and choose the local minimums among critical points. Due to the difficulty of obtaining analytical solutions from transcendental equations $\frac{dU(\mathbf{q})}{d\mathbf{q}} = \mathbf{0}$, quasi-Newton methods [Boyd and Vandenberghe, 2004] will be used in this chapter. The following proposition narrows down the search space when implementing numerical approaches.

Proposition 4.2 *Given a negative Gaussian potential well and a negative Gaussian potential trench with sources* $\mathbf{q}^{s}$ *and* $c^{s}(\mathbf{q}) = 0$*, respectively, in* D^3*, the local minimums are always confined within*

- *the line segment* $\overline{\mathbf{q}^{t}\mathbf{q}^{s}}$ *for a linear trench; or*
- *the curve* $\widetilde{\mathbf{q}^{t}\mathbf{q}^{s}}$ *for a circular arc trench*

where $\mathbf{q}^{t}$ *is the tangent point to equipotential surfaces of the well, and* $\mathbf{q}^{t}$ *satisfies* $c^{s}(\mathbf{q}^{t}) = 0$.

Proof: For a linear constraint, first we prove that the tac-locus $\overline{\mathbf{q}^{t}\mathbf{q}^{s}}$ is a line segment. The equipotential surface of a Gaussian potential well can be represented as an ellipsoid $\frac{x^2}{a^2} + \frac{y^2}{b^2} + \frac{z^2}{c^2} = 1$. By assuming the tangent point $\mathbf{q}^{t}$ as (x_0, y_0, z_0) on the ellipsoid, the tangent plane at (x_0, y_0, z_0) can be represented as $\frac{x_0 x}{a^2} + \frac{y_0 y}{b^2} + \frac{z_0 z}{c^2} = 1$, which is also one of the equipotential surfaces of the Gaussian potential trench. We scale the ellipsoid to find if the tac-locus is a line segment. Before scaling, the tangent plane equation is $\frac{x_0 x}{a^2} + \frac{y_0 y}{b^2} + \frac{z_0 z}{c^2} = 1$. For any given $\lambda > 0$, the ellipsoid can be scaled as $\frac{x^2}{a^2\lambda^2} + \frac{y^2}{b^2\lambda^2} + \frac{z^2}{c^2\lambda^2} = 1$. The tangent point on the ellipse is also scaled to $(\lambda x_0, \lambda y_0, \lambda z_0)$, where the tangent plane equation is $\frac{x_0}{a^2\lambda}x + \frac{y_0}{b^2\lambda}y + \frac{z_0}{c^2\lambda}z = 1$. While scaling, the normal vectors of the tangent planes remain invariant. Due to the uniqueness of the tangent point, the tac-locus is a line when the ellipsoid is scaled, and the tangent plane moves in parallel.

Next, we prove that the local minimums are always confined within the line segment $\overline{\mathbf{q}^{t}\mathbf{q}^{s}}$. The data space can be separated into three parts by two planes

$$c^{s}(\mathbf{q}) = 0 \tag{4.19}$$

$$c^{s}(\mathbf{q}) = \delta \text{ satisfying } c^{s}(\mathbf{q}^{s}) = \delta \tag{4.20}$$

where δ is the translational distance. The local minimums exist only between those two planes, as for any point beyond the region, we can always find a point with less potential at $\mathbf{q}^{s}$ or $\mathbf{q}^{t}$ with the tangent plane $c^{s}(\mathbf{q}) = 0$. For any point on each equipotential plane between $c^{s}(\mathbf{q}) = 0$ and $c^{s}(\mathbf{q}) = \delta$, we can always find a point with less potential at the tangent point $\mathbf{q}^{t}$.

For an arc constraint, although the tac-locus $\widetilde{\mathbf{q}^{t}\mathbf{q}^{s}}$ is not a line segment, the similar proposition can be proven on the basis of the previous thoughts. □

Now let us discuss the implications of local minimums with regard to linear constraints, as most road segments are linear in digital maps. As shown in Figure 4.2(a)-(c), in a 3D data space with a point potential well source $\mathbf{q}^{s} = (x^{s}, y^{s}, z^{s})^{T}$ and a potential trench source $c^{s}(\mathbf{q}) = p_a x + p_b y + p_c z + p_d = 0$, we can always find a set of new Cartesian coordinates to eliminate the term p_c by using a rotation matrix $\mathbf{R}$, with potential field shape invariant. Herein we assume

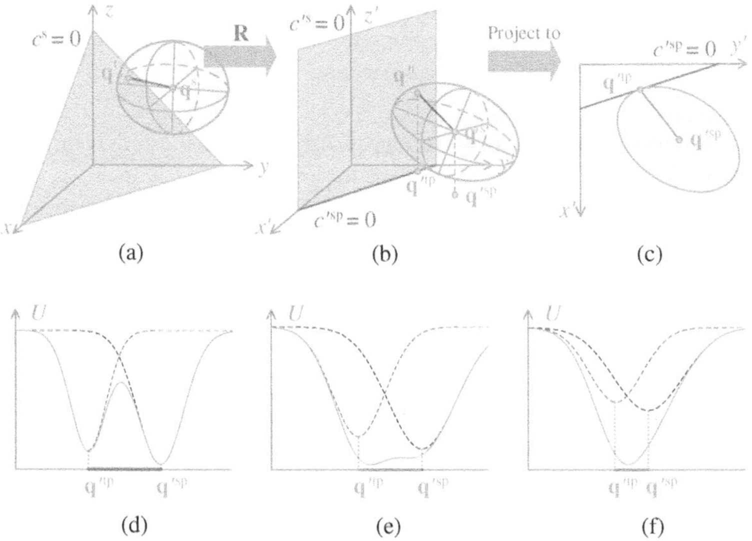

Figure 4.2 Several situations with minimums in the resultant potential field. As the equipotential ellipsoids and planes have been projected to a 2D plane, the curves in (d), (e), and (f) are generally skewed. Also note that the local minimums are always confined within line segment $\overline{\mathbf{q}'^{t}\mathbf{q}'^{s}}$.

the sources after rotation are $\mathbf{q}'^{s} = (x'^{s}, y'^{s}, z'^{s})^T$, and $c'^{s}(\mathbf{q}') = p'_{a}x' + p'_{b}y' + p'_{d} = 0$. According to Proposition 4.2, we consider the potential on the projection of line segment $\overline{\mathbf{q}'^{t}\mathbf{q}'^{s}}$: $\overline{\mathbf{q}'^{tp}\mathbf{q}'^{sp}}$, where $\mathbf{q}'^{t}$ is the tangent point between $c'^{s}(\mathbf{q}') = 0$ and the equipotential ellipsoid. The potential on $\overline{\mathbf{q}'^{tp}\mathbf{q}'^{sp}}$ is associated with the parameters of potential construction functions and can be classified into three categories demonstrated in Figure 4.2(d)-(f).

Figure 4.2(d) shows a bimodal curve where there are two local minimums. This happens when the distance between $\mathbf{q}^{s}$ and $c^{s}(\mathbf{q}) = 0$ is much larger than the affected range of sources. Figure 4.2(e) shows the transition between (d) and (f), with two critical points but only one local minimum. Figure 4.2(f) shows one local minimum. This happens when sources are close or the influence of fields is sustained over a long distance.

In numerical optimizing methods, the initial searching point plays a major role in the final result. By setting the starting point as the potential well source $\mathbf{q}^{s}$, there are several properties concerning minimums, which are as follows:

Property 4.1 The number of local minimums is influenced by the distance between sources and source covariances. The total potential field tends to be unimodal with smaller distances and larger source covariances.

Property 4.2 In the case described in Figure 4.2(d), the positioning result is close to $\mathbf{q}'^{s}$. This means that when the measurement is in discord with the constraints, the fusion scheme tends to trust the measurement.

Property 4.3 In the case described in Figure 4.2(f), the positioning result falls between $\mathbf{q}'^{s}$ and $\mathbf{q}'^{t}$. This means that when the measurement is near the constraints, the constraints and measurements are combined to provide a more accurate position.

In vehicle positioning problems, it is assumed that the vehicle is always following the road constraints. However, sometimes the assumption is invalid when the venicle temporarily leaves the road or due to obsolete road maps. The fusion frame based on a potential field considers this circumstance inherently. According to Properties 4.1, 4.2 and 4.3, when the measurement deviates from the constraints, if the measurement is accurate (with a small covariance matrix), the positioning result tends to follow Figure 4.2(d), where the constraint has low reliability. Otherwise, if the measurement is inaccurate (with a large covariance matrix), the result tends to follow Figure 4.2(f), and the measurement is matched to the constraint.

4.2 Potential-Function-Based Fusion for Vehicle Positioning

The localization system consists of several data sources with different reference frames. These reference frames are defined here [Wei et al., 2013]: VO uses a right-handed camera frame whose origin is at the center of the left camera, the x_c axis points to the center of the right camera, and the y_c axis points downward along the image plane. GPS and digital maps output data in the WGS-84 coordinate system with longitude and latitude (altitude is only available for GPS). All the coordinates in the camera frame and WGS-84 frame are converted or projected to local Cartesian coordinates in the East-North-Up (ENU) working frame O_{xyz}.

4.2.1 Information Sources and Sensors

Visual Odometry Essentially, VO is a dead-reckoning localization approach that calculates the localization recursively. In a stereo system, the problem can be formulated as follows [Scaramuzza and Fraundorfer, 2011]: given a homogeneous initial camera pose $\mathbf{C}_0 \in \mathbb{R}^{4\times1}$, a set of camera images $I_{l,0:n} = \{I_0, \dots, I_n\}$, $I_{r,0:n} = \{I_0, \dots, I_n\}$, and the calibration data of the camera, compute the relative transformation matrices $\mathbf{T}_{k,k-1} \in \mathbb{R}^{4\times4}$ that satisfy the relationship between camera poses $\mathbf{C}_k$ at different times

$$\mathbf{C}_k = \mathbf{T}_{k,k-1}\mathbf{C}_{k-1} = \begin{bmatrix} \mathbf{R}_{k,k-1} & \mathbf{t}_{k,k-1} \\ 0 & 1 \end{bmatrix} \mathbf{C}_{k-1} \tag{4.21}$$

where $k = 1, \dots, n$ denotes a discrete time instant, $\mathbf{R}_{k,k-1} \in SO(3)$ is the rotation matrix, and $\mathbf{t}_{k,k-1} \in \mathbb{R}^{3\times1}$ is the translation vector. In this chapter, the VO algorithm in Geiger et al. (2011) is implemented.

Due to the recursive nature of VO drift error [Jiang et al., 2010], we model the covariance matrix $\mathbf{D}_{\mathrm{VO}}$ as

$$\mathbf{D}_{\mathrm{VO}} = \mathbf{D}_0\eta^s \tag{4.22}$$

where $\mathbf{D}_0$ is the initial covariance matrix, $\eta > 1$ is a user-defined parameter, and s is the current displacement measured by VO.

Digital Maps Herein, digital maps providing road constraints are used to create potential trenches. Digital maps are generally represented as nodes and ways, where each node consists of its ID, latitude, and longitude and each way is an ordered list of nodes. After projection, the digital maps are flattened, with nodes and ways represented in 2D Cartesian space without a z axis. A general linear road constraint in a 2D map is a plane in 3D data space and can be represented as $c^s(\mathbf{q}) = p_{\mathrm{a}}x + p_{\mathrm{b}}y + p_{\mathrm{d}} = 0$, where $p_{\mathrm{a}}, p_{\mathrm{b}}, p_{\mathrm{c}}$ are shaping parameters. The potential generated by $c^s(\mathbf{q}) = 0$ at point $\mathbf{q}(x, y, z)$ can be calculated from Eqs. (4.16) and (4.17), where

$$d_{\mathrm{t}} = \frac{|p_{\mathrm{a}}x + p_{\mathrm{b}}y + p_{\mathrm{d}}|}{\sqrt{p_{\mathrm{a}}^2 + p_{\mathrm{b}}^2}} \tag{4.23}$$

Sometimes the road cannot be simply modelled as linear segments. Compared to linear models, arc models keep positioning results smoother because of less modeling error and fewer switches between constraints. An arc model with center $(x_{\mathrm{c}}, y_{\mathrm{c}})$ and radius r is represented as $c^s(\mathbf{q}) = (x - x_{\mathrm{c}})^2 + (y - y_{\mathrm{c}})^2 - r^2 = 0$. Similarly, the potential at point $\mathbf{q}(x, y, z)$ can be calculated from Eqs. (4.16) and (4.17), where

$$d_{\mathrm{t}} = |\sqrt{(x - x_{\mathrm{c}})^2 + (y - y_{\mathrm{c}})^2} - r| \tag{4.24}$$

The confidence coefficient σ in Eq. (4.17) is selected according to the road conditions and experience. In general, for a wider road, σ should be larger to indicate less confidence that the vehicle is following the road centerline.

GPS Like digital maps, GPS outputs the vehicle's position with latitude, longitude, and altitude. The coordinate transformation makes it compatible with local ENU coordinates. The covariance of GPS measurement $\mathbf{D}_{\mathrm{GPS}}$ can be obtained from

National Marine Electronics Association (NMEA) messages GST, which provides position error statistics such as standard deviations for latitude, longitude, and height [Wei et al., 2013].

4.2.2 Potential Representation

With multiple information sources as described, based on Eq. (4.18), the total potential field in data space is represented with Eq. (4.25). The gradient $\nabla U(\mathbf{q})$ can be

$$
\begin{aligned}
U(\mathbf{q}) &= \alpha_{\mathrm{w}} f_{\mathrm{w}}(\mathbf{d}_{\mathrm{w}}) + \alpha_{\mathrm{t}} f_{\mathrm{t}}(d_{\mathrm{t}}) \\
&= -\frac{\alpha_{\mathrm{w}}}{(2\pi)^{\frac{3}{2}} |\mathbf{D}|^{\frac{1}{2}}} \exp\left(-\frac{1}{2}\mathbf{d}_{\mathrm{w}}^{T}\mathbf{D}^{-1}\mathbf{d}_{\mathrm{w}}\right) - \frac{\alpha_{\mathrm{t}}}{\sigma\sqrt{2\pi}} \exp\left(-\frac{d_{\mathrm{t}}^{2}}{2\sigma^{2}}\right) \\
&= \begin{cases}
-\dfrac{\alpha_{\mathrm{w}}}{(2\pi)^{\frac{3}{2}} |\mathbf{D}|^{\frac{1}{2}}} \exp\left(-\dfrac{1}{2}(\mathbf{q}-\mathbf{q}^{\mathrm{s}})^{T}\mathbf{D}^{-1}(\mathbf{q}-\mathbf{q}^{\mathrm{s}})\right) \\
\quad -\dfrac{\alpha_{\mathrm{t}}}{\sigma\sqrt{2\pi}} \exp\left(-\dfrac{(p_{\mathrm{a}}x + p_{\mathrm{b}}y + p_{\mathrm{d}})^{2}}{2\sigma^{2}(p_{\mathrm{a}}^{2} + p_{\mathrm{b}}^{2})}\right), & \text{for linear trench} \\
-\dfrac{\alpha_{\mathrm{w}}}{(2\pi)^{\frac{3}{2}} |\mathbf{D}|^{\frac{1}{2}}} \exp\left(-\dfrac{1}{2}(\mathbf{q}-\mathbf{q}^{\mathrm{s}})^{T}\mathbf{D}^{-1}(\mathbf{q}-\mathbf{q}^{\mathrm{s}})\right) \\
\quad -\dfrac{\alpha_{\mathrm{t}}}{\sigma\sqrt{2\pi}} \exp\left(-\dfrac{|\sqrt{(x-x_{\mathrm{c}})^{2} + (y-y_{\mathrm{c}})^{2}} - r|^{2}}{2\sigma^{2}}\right), & \text{for arc trench.}
\end{cases}
\end{aligned}
\tag{4.25}
$$

obtained as

$$\nabla U(\mathbf{q}) = \nabla U_{\mathrm{w}}(\mathbf{q}) + \nabla U_{\mathrm{t}}(\mathbf{q}) \tag{4.26}$$

4.2.3 Road-Switching Strategy

The correct selection of road constraints affects positioning performance. As mentioned earlier, a scheme is proposed for road switching to find the most appropriate road constraint at all times. The switching strategy includes (i) a switching condition, and (ii) new constraint determination. As is the case with arc constraints, switching is considered when the distance between the sources of potential well and trench $d_{\mathrm{t}} > \epsilon$, where ϵ is a parameter defined according to constraint covariance σ, the density of the road network, and measurement accuracy. Like linear segments in most cases, we define the heading difference $\Delta\theta = |\theta_{\mathrm{rd}} - \theta_{\mathrm{m}}|$, where θ_{rd} denotes the heading of the road constraint, and θ_{m} is the heading obtained from two consecutive measurement positions [Jagadeesh et al., 2004]. The road-switching scheme is trigged once $\Delta\theta > 10°$ and $d_{\mathrm{t}} > \epsilon$.

Algorithm 2 Road switching strategy

Input: Linear candidates $C^s = \{c_1^s(\mathbf{q}) = 0, \cdots, c_n^s(\mathbf{q}) = 0\}$, arc candidates $C_{arc}^s = \{c_{arc1}^s(\mathbf{q}) = 0, \cdots, c_{arcm}^s(\mathbf{q}) = 0\}$, current measurement $\mathbf{q}^s$, current road c_{bfr}^s, heading difference $\Delta\theta$, distance between measurement and road segment d_t

Output: Road segment after switching c_{aft}^s

  1:   $i \leftarrow 1$
  2:   flag $\leftarrow 0$
  3:   $d_{min} \leftarrow 50$
  4:   $\Delta\theta_{min} \leftarrow 60$
  5:   **if** switching condition trigged **then**
  6:     **while** $i \leq m$ **do**
  7:       Calculate the distance d_{arci} between $\mathbf{q}^s$ and each arc in C_{arc}^s
  8:       **if** $d_{arci} < d_{min}$ **then**
  9:         $c_{aft}^s \leftarrow c_{arci}^s(\mathbf{q}) = 0$
10:         flag $\leftarrow 1$
11:         $d_{min} \leftarrow d_{arci}$
12:       **end if**
13:       $i \leftarrow i + 1$
14:     **end while**
15:     **if** c_{aft}^s is not empty **then**
16:       **goto** 29
17:     **end if**
18:     $i \leftarrow 1$
19:     **while** $i \leq n$ **do**
20:       Calculate the heading difference $\Delta\theta_i$ for each segment in C^s
21:       Calculate the distance d_{ti} between $\mathbf{q}^s$ and each segment in C^s
22:       **if** $d_{ti} < 50$ and $\Delta\theta_i < \Delta\theta_{min}$ **then**
23:         $c_{aft}^s \leftarrow c_i^s(\mathbf{q}) = 0$
24:         flag $\leftarrow 0$
25:       **end if**
26:       $i \leftarrow i + 1$
27:     **end while**
28:   **end if**
29:   **return** c_{aft}^s

After the switching condition is trigged, the arc constraint with minimum d_t has priority to be the new constraint. Otherwise, the linear road segment around the measurement with the minimum $\Delta\theta$ is selected as the new constraint.

The entire road-switching process is demonstrated in Algorithm 2.

4.3 Experimental Results

The summary of the test sequences is listed in Table 4.1. The proposed fusion approach is tested with data provided by the Karlsruhe Dataset (sequences 1-3) [Geiger et al., 2011a], the Málaga Stereo and Laser Urban Data Set (sequences 4-5) [Blanco-Claraco et al., 2014], and our data (sequences 6). In the first five datasets, the true position is provided by a combined GPS/IMU system, and the information is fused according to road constraints and VO data. In our data, the true position is provided by a Trimble DSM12/212 DGPS. A low-cost GPS (ublox EVK-M8F), VO equipped with a Bumblebee2 BB2-08S2C stereo camera, and road constraints are combined to obtain the estimated position. All self-generated road constraints in the experiments are created from OpenStreetMap, while the map-matching results are obtained from the Google Maps Roads API.

4.3.1 Quantitative Results

The evaluations are undertaken with the pure VO, map matching (MM), and the potential field (PF) method, respectively. The performance of these methods is demonstrated in Table 4.2. As only 2D maps are provided in MM, the vertical mean error and vertical standard deviation are not included in the statistics. In the experiments, the parameters in the potential construction functions are set as $\alpha_t = 10$, $\alpha_w = 200$, and trench covariance $\sigma = 20$; initial VO covariance is set as $\mathbf{D}_0 = \text{diag}(5, 5, 5)$ for sequences 1-5 and $\mathbf{D}_0 = \text{diag}(20, 20, 20)$ for sequence 6, with different increasing rates η from 1.005 to 1.01.

The evaluation indexes include total error, mean error, and standard deviation. The total error is calculated from the difference in traveled distance compared to the ground truth. The mean error and standard deviation are obtained by

Table 4.1 Test sequences.

No.	Length (m)	Information source			Description
		VO	GPS	Road maps	
1	225.66	✓		✓	Short path with turns
2	271.89	✓		✓	Mostly straight path with backward movement and small turns
3	237.09	✓		✓	Path with loops
4	594.17	✓		✓	Newly built road with an obsolete road map
5	1389.23	✓		✓	Block loop closure
6	1326.50	✓	✓	✓	Self-recorded path at low speed

Table 4.2 Positioning results with different methods.

No.	Method	Error		
		Total%	$Mean_{xy}$ (m)	Std_{xy} (m)
1	VO	3.49	4.24	1.34
	VO+MM	11.35	3.68	1.97
	PF	6.91	3.78	1.32
2	VO	5.07	9.97	3.39
	VO+MM	14.90	10.47	3.37
	PF	8.22	9.96	3.36
3	VO	3.38	2.35	0.93
	VO+MM	8.86	3.42	1.13
	PF	2.67	2.30	0.94
4	VO	4.27	23.46	28.35
	VO+MM	10.16	34.71	16.84
	PF	4.36	23.45	16.51
5	VO	1.90	14.32	12.87
	VO+MM	3.17	11.91	9.12
	PF	4.28	12.40	9.07
6	VO	46.47	20.69	4.62
	GPS	39.55	8.58	9.44
	PF	41.59	7.53	5.07

evaluating the positioning results for each frame. The VO accuracy is relatively low when measuring straight movements and short distances compared to turns, loops, and larger-scale environments. The performance of MM is influenced by the vehicle's path: driving close to the road centerline leads to a smaller mean error, especially for wide roads, as most roads are modeled using centerlines in digital maps. According to the test results, the proposed fusion approach shows adaptability and robustness, as it achieves a good balance among the evaluation indexes. The mean error and standard deviation approach the best results measured by the various methods given different road conditions. It must be pointed out that in sequence 6, the large total error is due to the different sampling rates of DGPS and other sensors. The path obtained from DGPS reflects the actual moving trajectory of the vehicle, and it is not as smooth as other paths. The DGPS measurements are down-sampled to 1 Hz to make comparisons with the fusion results.

4.3.2 Qualitative Evaluation

In this section, we mainly focus on two indexes of positioning results: small-scale sensitivity and large-scale drift. In other words, movement in all directions should be detected in a short distance, while the drifting error should be reduced during a long drive.

It is difficult for the MM method to reflect the local movement of the vehicle, especially in the vertical direction of the road segment, as movement in this direction is regarded as invalid along the road. In sequence 2, the vehicle reverses while traveling. As can be seen in Figure 4.3, the PF-based method utilizes VO measurements in small-range movement, and the positioning error is bounded by restraining VO drifting. In sequence 3, a looping trajectory is tested. Figure 4.4 shows the poor continuity of the MM result. The proposed PF method can balance trajectory smoothness and localization mean error by adjusting the covariances of the potential functions. Due to the pace of urban construction, digital maps sometimes are not updated promptly. Sequence 4 considers the circumstance of driving along a newly built road. Figure 4.5 and Figure 4.6 illustrate that our approach outperforms MM when the vehicle is on a new path from frame 40 to 72, and it also reduces error compared to VO after frame 82.

Pure VO suffers from drifting problems along with distance. In sequence 5, a block loop closure path is tested, with the results shown in Figure 4.7 and Figure 4.8. The PF-based approach produces a robust result compared to the pure VO method. The initial VO measurements with little drifting error tend to

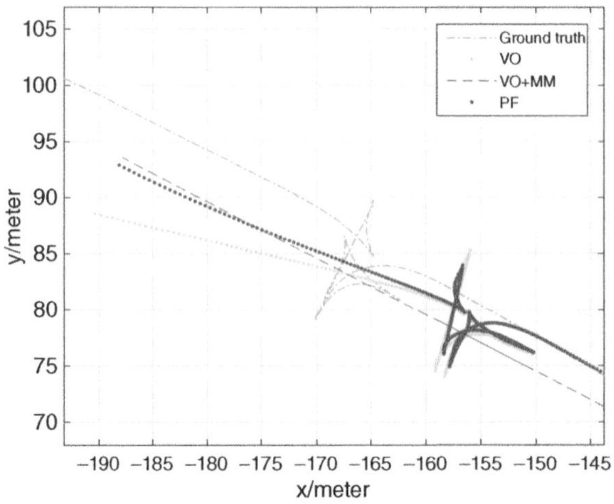

Figure 4.3 Partial positioning results of sequence 2.

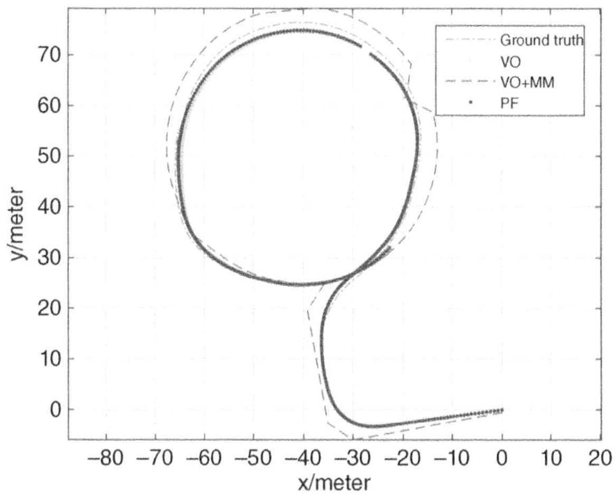

Figure 4.4 Positioning error of sequence 3.

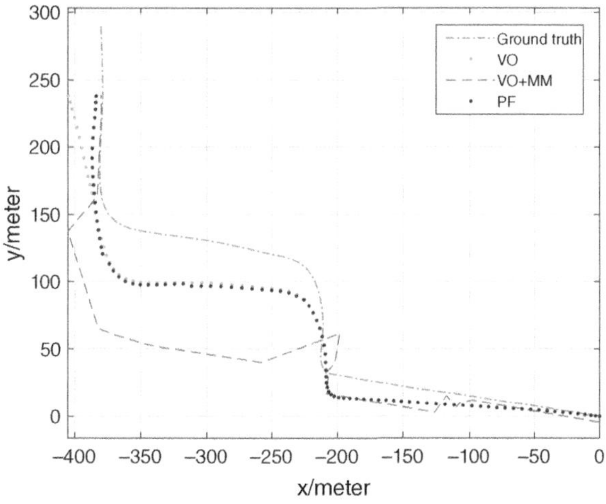

Figure 4.5 Positioning results of sequence 4.

be trusted, while the fusion results are similar to MM positions at the end. As shown in Figure 4.9 and Figure 4.10, in sequence 6, the low-cost GPS provides an unstable path that deviates from the ground truth when the vehicle is in urban canyon environments. Moreover, the VO path drifts such that only the geometry outline of the ground truth is presented. After PF-based fusion, the estimated results show greatly improved positioning performance.

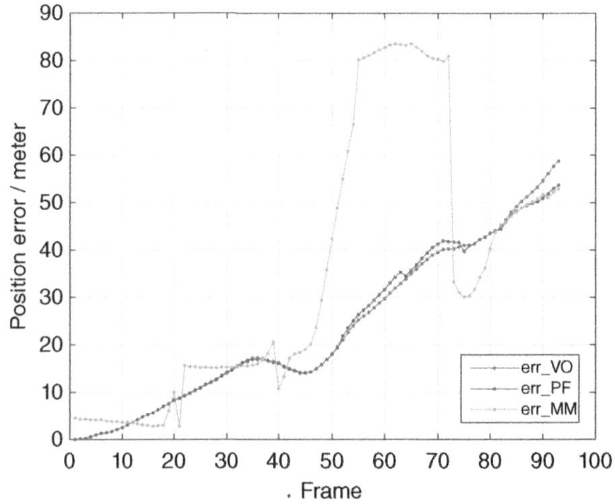

Figure 4.6 Positioning error of sequence 4.

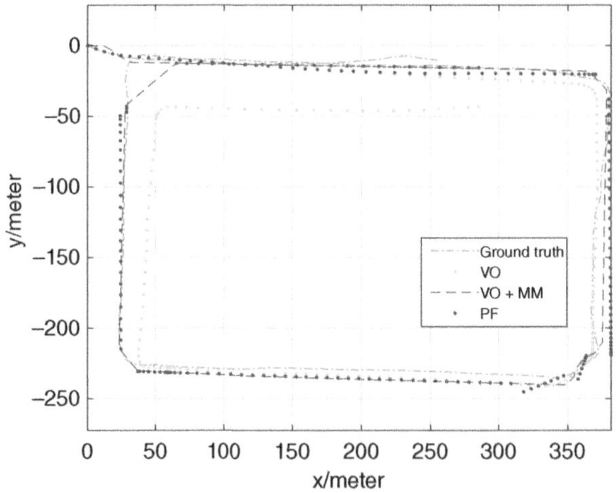

Figure 4.7 Positioning results of sequence 5.

The proposed approach is implemented in Matlab, and all the experiments are conducted on a mobile workstation with an i7-4710MQ/2.5 GHz processor. It takes 16 ms per frame for minimum searching and 2 ms per frame for road switching.

Our work has some limitations. As we have not taken the dynamic model of the vehicle into consideration, the smoothness of the trajectory is not satisfying

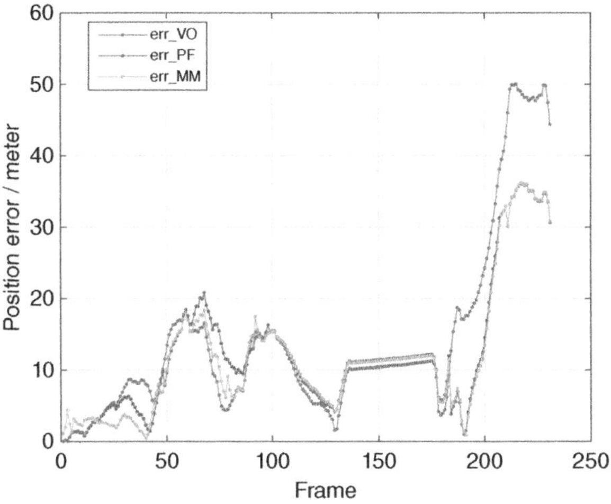

Figure 4.8 Positioning error of sequence 5.

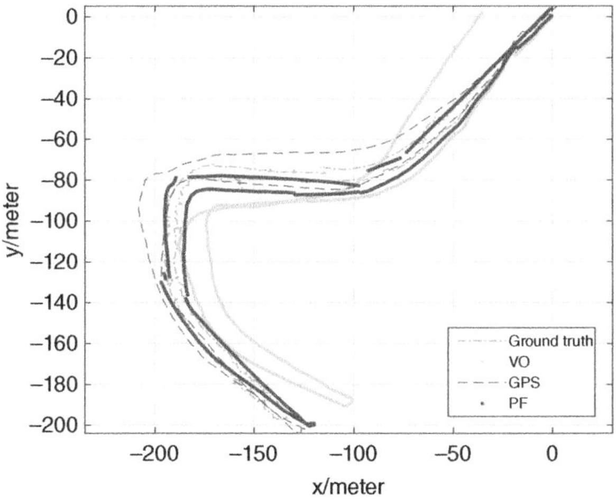

Figure 4.9 Positioning results of sequence 6.

under specific conditions, which will have a negative effect on user experience even though the average positioning error is small. In particular, the trajectory is not smooth when the vehicle's path consists of short road segments. By creating the dynamic potential field, the dynamic model of the vehicle could be combined within the fusion frame such that trajectory smoothness could be improved.

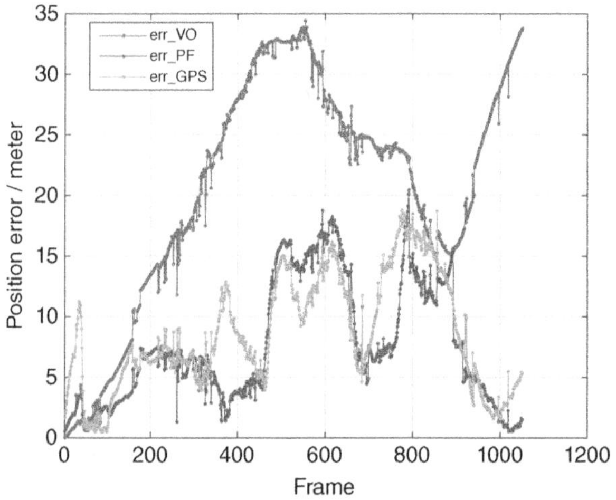

Figure 4.10 Positioning error of sequence 6.

4.4 Conclusion

In this chapter, we have presented a potential-field-based fusion approach for vehicle positioning with a low-cost GPS, VO, and digital maps. By creating potential wells and potential trenches separately, the proposed approach intuitively combines multi-source information, including position measurements and road constraints, without additional map matching. The position is then estimated by searching for the minimum of the combined potential field. Although it is almost impossible to obtain the analytic minimum from transcendental equations, the local minimums are proven to be confined such that the numerical searching space is greatly reduced. The experiment results show that the potential-field-based approach combines the advantages of multiple types of data and provides accurate and robust localization when high-precision GPS is not available.

For future work, efforts can be made using more types of sensors. In addition, we would like to explore the creation of a dynamic potential field by considering system state equations and measurement equations as a part of the fusion frame.

5

Multi-Sensor Geometric Pose Estimation

As mentioned in Chapter 3, constraints may be of great help in improving pose estimation performance. Thanks to the construction of digital map databases such as Google Maps and OpenStreetMap (OSM), the most common constraints for ground vehicles are road constraints, which assume the vehicle stays on the road at all times. Unfortunately, none of these approaches considers constraints inherently. For filtering-based approaches, interior point likelihood maximization [Bell et al., 2009], projection methods [Simon and Chia, 2002], and many others [Simon, 2010] have been applied. For graph optimization, constraints can be added to the cost function such that minimizing the cost function leads to constraint satisfaction.

In Jiang et al. (2018), potential wells and potential trenches have been used to represent measurements and constraints such that vehicles can be localized with a low-cost GPS, visual odometry (VO), and a digital map. However, the previous work considers system states at a discrete time with no evolution, which makes it difficult to incorporate the state equation into the proposed framework. Furthermore, potential fields in Euclidean space limit the applications of data fusion since data may have different topological structures other than Euclidean. In this chapter, by representing measurements and constraints geometrically, we introduce a data fusion model using on-manifold dynamic potential fields (DPFs). The pose estimation problem is then formulated based on the proposed model, in which the posterior distributions can be obtained from Bayesian inference.

In Section 5.1, preliminary technical details are summarized to help understand Geometric Pose Estimation (Geo-PE), which is presented in Section 5.2. To facilitate the constrained pose estimation problem for ground vehicles, Section 5.3 proposes to integrate SVO, headings, and maps based on Geo-PE. Experimental configurations and results are shown in Section 5.4 and 5.5. Finally, Section 5.6 concludes the chapter.

Multimodal Perception and Secure State Estimation for Robotic Mobility Platforms, First Edition.
Xinghua Liu, Rui Jiang, Badong Chen, and Shuzhi Sam Ge.

5.1 Preliminaries

This section provides basic concepts from differential geometry to help understand geometric pose estimation. Some notations related to Lie group are listed in Table 5.1. Interested readers may refer to Chirikjian (2009, 2011) and Pennec (2004) for more details and formal definitions.

5.1.1 Distance on Riemannian Manifolds

Given a manifold M, we can further define a Riemannian metric, which is a smooth, positive-definite, non-degenerate bilinear map $\langle \cdot, \cdot \rangle_p : T_pM \times T_pM \to \mathbb{R}$ on the tangent space T_pM at each point $p \in M$. We call $(M, \langle \cdot, \cdot \rangle_p)$ a Riemannian manifold. The length of curve $\gamma(t)$ on $(M, \langle \cdot, \cdot \rangle_p)$ can be computed by

$$l(\gamma) = \int_a^b \left(\langle \dot{\gamma}(t), \dot{\gamma}(t) \rangle_{\gamma(t)} \right)^{\frac{1}{2}} \mathrm{d}t \tag{5.1}$$

where $\dot{\gamma}(t_0)$ denotes the tangent vector of $\gamma(t)$ at $t = t_0$.

The distance between two points p_1, p_2 on connected Riemannian manifold M can be defined as the minimum curve length connecting p_1 and p_2:

$$d(p_1, p_2) = \min_{\gamma} l(\gamma), \text{with } \gamma(0) = p_1, \gamma(1) = p_2. \tag{5.2}$$

The calculus of variations shows that geodesics lead to minimum length.[1] Given a point $p \in M$ and a tangent vector $\frac{\partial}{\partial t} \in T_pM$, the geodesic $\gamma(t)$ can be uniquely determined such that $\gamma(0) = p$ and $\dot{\gamma}(0) = \frac{\partial}{\partial t}$. In this chapter, by assuming M is geodesically complete, the Hopf-Rinow theorem states that there always exists $d(p_1, p_2)$ satisfying Eq. (5.2) for any $p_1, p_2 \in M$.

Table 5.1 Notations from Lie group and differential geometry.

$g \in G, p \in M$	An element in group G, a point on manifold M
T_pM	Tangent space of M at p
$\mathrm{SE}(m)$	Special Euclidean group in dimension m
$\mathrm{SO}(m)$	Special orthogonal group in dimension m
$\exp_p(\cdot) : T_pM \to M$	Exponential map from tangent space at p to manifold M
$\exp(\cdot) : \mathfrak{g} \to G$	Exponential map from Lie algebra $\mathfrak{g}$ to Lie group G
$\log_p(\cdot) : M \to T_pM$	Logarithmic map from manifold M to tangent space at p
$\log(\cdot) : G \to \mathfrak{g}$	Logarithmic map from Lie group G to Lie algebra $\mathfrak{g}$
$(\cdot)^\vee : \mathfrak{g} \to \mathbb{R}^m$	Linear isomorphism from Lie algebra to m-dimensional vector

1 Strictly, geodesic γ leads to minimum length if and only if there is no conjugate points along γ.

5.1.2 Probabilistic Distribution on Riemannian Manifolds

Herein, probabilities and statistics have been generalized on Riemannian manifolds, adapting to state spaces other than Euclidean.

Definition 5.1 ***(Random primitive)*** Let $(\Omega, \mathcal{B}, \Pr)$ be a probability space with the Borelian tribe. A random primitive on the Riemannian manifold M is a Borelian function $X : \Omega \to M$.

Definition 5.2 ***(Probability density function, PDF)*** Let $\mathcal{A}$ be the Borelian tribe of M. The random primitive X has a probability density function $\mathbb{P}$ if $\forall\, \mathcal{X} \in \mathcal{A}$ and $p \in M$,

$$\Pr(X \in \mathcal{X}) = \int_{\mathcal{X}} \mathbb{P}(p) \mathrm{d}M(p) \tag{5.3}$$

$$\text{and } \Pr(X \in M) = \int_{M} \mathbb{P}(p) \mathrm{d}M(p) = 1 \tag{5.4}$$

where $\mathbb{P} : M \to \mathbb{R}$ is a non-negative and integrable function; $\mathrm{d}M$ denotes a measure on M.

Definition 5.3 ***(Expectation of a function)*** Let $\varphi(p) : M \to \mathbb{R}$ be a Borelian function on M and X be a random primitive with PDF $\mathbb{P}$; then $\varphi(X) : \Omega \to \mathbb{R}$ is a random variable for which we can compute the expectation from

$$\mathbf{E}[\varphi(X)] = \int_{M} \varphi(p) \mathbb{P}(p) \mathrm{d}M(p) \tag{5.5}$$

Definition 5.4 ***(Variance of a random primitive)*** Let X be a random primitive with PDF $\mathbb{P}$. Given a fixed point $p_0 \in M$, the variance of X is defined as

$$\sigma_X^2(p_0) = \mathbf{E}\left[d(p_0, X)^2\right] \tag{5.6}$$

where d denotes the distance between two points.

Definition 5.5 ***(Expectation of a random primitive)*** Followed by Definition 5.4, we have the expectation notation

$$\mathbb{E}[X] = \underset{p_0 \in M}{\arg\min}\, \mathbf{E}\left[d(p_0, X)^2\right] \tag{5.7}$$

Note that the existence and uniqueness of $\mathbb{E}[X]$ are not guaranteed due to the optimization-based definition. The existence and uniqueness have been further discussed in Karcher (1977) and Kendall (1990), where local minima are also covered as "Karcher means."

Definition 5.6 ***(Covariance matrix of a random primitive)*** The covariance matrix relative $\bar{X} \in \mathbb{E}[X]$ is defined as

$$\mathrm{Cov}_{\bar{X}}(X) = \mathbf{E}\left[\left[\mathbf{d}(\bar{X}, X)\right]\left[\mathbf{d}(\bar{X}, X)\right]^{\mathsf{T}}\right] \tag{5.8}$$

where $\mathbf{d}(\bar{X}, X)$ denotes the distance vector from $\bar{X}$ to X in Definition 5.11.

5.2 Geometric Pose Estimation Using Dynamic Potential Fields

5.2.1 State Space and Measurement Space

By enabling different topological structures on state variables, the state space can be generalized from Euclidean to a manifold M, where a state at discrete time k is denoted as a point $\mathbf{x}_k \in M$. Similarly, we introduce the measurement space N, a manifold on which each point $\mathbf{z}_k \in N$ denotes a particular measurement at k. The ideal state equation and output equation can be deemed as smooth mappings $\mathbf{f}(\cdot, \mathbf{u}) : M \to M$ and $\mathbf{h}(\cdot) : M \to N$, respectively.[2] Then the state equation and output equation can be represented as compositions

$$\mathbf{f}_{\mathbf{w}} \circ \mathbf{f} : M \to M \tag{5.9}$$

$$\mathbf{h}_{\mathbf{v}} \circ \mathbf{h} : M \to N \tag{5.10}$$

where $\mathbf{f}_{\mathbf{w}} : M \to M$ and $\mathbf{h}_{\mathbf{v}} : N \to N$ introduce process noise and measurement noise, respectively.

Under the geometric perspective, we then have the following definitions [Hermann and Krener, 1977, Albertini and D'Alessandro, 1996] and proposition:

Definition 5.7 A system is *invertible* if and only if for every admissible control input $\mathbf{u}$, $\mathbf{f}(\cdot, \mathbf{u})$ is a diffeomorphism.

Proposition 5.1 *In the pose estimation problem where state space* $M = \mathrm{SE}(3)$, *the system is invertible.*

Proof: We may represent $\mathbf{u}_{k-1}$ as a transformation matrix $\mathtt{T}_{k-1}$ such that the state equation is $\mathtt{C}_k = \mathtt{C}_{k-1}\mathtt{T}_{k-1}$. Mapping $\mathbf{f}(\cdot, \mathtt{T}_{k-1}) : \mathtt{C}_{k-1} \mapsto \mathtt{C}_k$ is obviously a diffeomorphism for any $\mathtt{T}_{k-1} \in \mathrm{SE}(3)$. □

---

2 Geometrically, points on manifolds are independent of coordinate systems. Herein, with a slight abuse of notation, $\mathbf{f}(\cdot, \mathbf{u})$, and $\mathbf{h}(\cdot)$ are used to represent mappings between manifolds; $\mathbf{x}_k$ and $\mathbf{z}_k$ denote points in state space and measurement space.

Definition 5.8 Two states $\mathbf{x}_0$, $\mathbf{x}'_0 \in M$ are *indistinguishable* (written as $\mathbf{x}_0 I \mathbf{x}'_0$) if for each admissible control sequence $\mathbf{u}_0, \cdots, \mathbf{u}_j$, where j is any integer index satisfying $j > 0$, we have $\mathbf{h}(\mathbf{x}_0) = \mathbf{h}(\mathbf{x}'_0)$ and

$$\mathbf{h}\left(\mathbf{f}(\cdot, \mathbf{u}_{j-1}) \circ \cdots \circ \mathbf{f}(\mathbf{x}_0, \mathbf{u}_0)\right) = \mathbf{h}\left(\mathbf{f}(\cdot, \mathbf{u}_{j-1}) \circ \cdots \circ \mathbf{f}(\mathbf{x}'_0, \mathbf{u}_0)\right)$$

where the ring operator $\circ$ denotes composition. Let $U \subset M$ and $\mathbf{x}_0, \mathbf{x}'_0 \in U$. Two states $\mathbf{x}_0$, $\mathbf{x}'_0$ are said to be U-indistinguishable (written as $\mathbf{x}_0 I_U \mathbf{x}'_0$) if $\mathbf{x}_0 I \mathbf{x}'_0$ and

$$\mathbf{x}_j, \mathbf{x}'_j \in U, \forall j \geq 0 \tag{5.11}$$

where $\mathbf{x}_j = \mathbf{f}(\mathbf{x}_{j-1}, \mathbf{u}_{j-1})$, $\mathbf{x}'_j = \mathbf{f}(\mathbf{x}'_{j-1}, \mathbf{u}_{j-1})$.

Definition 5.9 A state $\mathbf{x}_0$ is *observable* if for each $\mathbf{x}'_0 \in M$, $\mathbf{x}_0 I \mathbf{x}'_0 \Rightarrow \mathbf{x}_0 = \mathbf{x}'_0$. A state $\mathbf{x}_0$ is *locally observable* if for every open neighborhood U of $\mathbf{x}_0$, $\mathbf{x}_0 I_U \mathbf{x}'_0 \Rightarrow \mathbf{x}_0 = \mathbf{x}'_0$. A system is (locally) observable if all states $\mathbf{x} \in M$ are (locally) observable.

Proposition 5.2 *By assuming noise as constant at each discrete time, given an invertible system as in Eqs. (1.1) and (7.2),*

1. *Mapping* $\mathbf{f}_\mathbf{w} \circ \mathbf{f}(\cdot, \mathbf{u})$ *is one-to-one and onto for any admissible control input* $\mathbf{u}$.
2. *If the system is locally observable, mapping* $\mathbf{h}_\mathbf{v} \circ \mathbf{h}(\cdot)$ *is one-to-one.*

Proof: Since noise is assumed constant, both $\mathbf{f}_\mathbf{w}$ and $\mathbf{h}_\mathbf{v}$ are diffeomorphisms.

1. Obviously, according to Definition 5.7, the diffeomorphism of $\mathbf{f}(\cdot, \mathbf{u})$ ensures that $\mathbf{f}_\mathbf{w} \circ \mathbf{f}(\cdot, \mathbf{u})$ is also one-to-one and onto.
2. As shown in Figure 5.1, assume that $\mathbf{h}_\mathbf{v} \circ \mathbf{h}(\cdot)$ is not one-to-one. Then $\exists \mathbf{x}_0$, $\mathbf{x}'_0 \in M$ such that $\mathbf{h}_\mathbf{v} \circ \mathbf{h}(\mathbf{x}_0) = \mathbf{h}_\mathbf{v} \circ \mathbf{h}(\mathbf{x}'_0)$. On the other hand, we can always find an open neighborhood that does not cover $\mathbf{h}_\mathbf{v} \circ \mathbf{h}(\mathbf{f}_\mathbf{w} \circ \mathbf{f}(\mathbf{x}_0, \mathbf{u}))$ nor

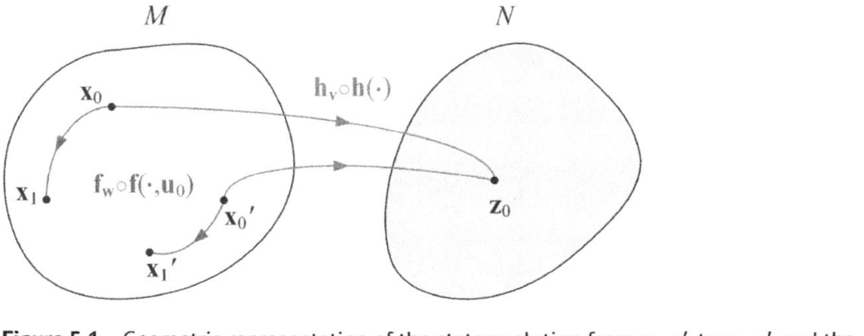

Figure 5.1 Geometric representation of the state evolution from $\mathbf{x}_0$, $\mathbf{x}'_0$ to $\mathbf{x}_1$, $\mathbf{x}'_1$ and the output equation. The control input is not shown, as our main concerns are state transitions and measurements.

$\mathbf{h_v} \circ \mathbf{h}(\mathbf{f_w} \circ \mathbf{f}(\mathbf{x}_0', \mathbf{u}))$ for any $\mathbf{u}$, as long as $\mathbf{f_w} \circ \mathbf{f}(\mathbf{x}_0, \mathbf{u}) \neq \mathbf{x}_0$ and $\mathbf{f_w} \circ \mathbf{f}(\mathbf{x}_0', \mathbf{u}) \neq \mathbf{x}_0'$. Since the system is locally observable, we have $\mathbf{h_v} \circ \mathbf{h}(\mathbf{x}_0) \neq \mathbf{h_v} \circ \mathbf{h}(\mathbf{x}_0')$, which is in contradiction to the assumption. Thus $\mathbf{h_v} \circ \mathbf{h}(\cdot)$ is one-to-one.

Remark 5.1 Mapping $\mathbf{h}(\cdot) : M \to N$ is not onto, since not all points on N have a preimage on M. An example would be $\mathbf{x} \in M = \mathbb{S}^1$, $\mathbf{h}(\mathbf{x}) = [\cos \mathbf{x}, \sin \mathbf{x}]^\top$, and $N = \mathbb{R}^2$.

5.2.2 Dynamic Potential Fields on Manifolds

In state space and measurement space, we represent states, measurements, and constraints with DPFs. Some definitions are introduced first.

Definition 5.10 ***(Potential function on a manifold)*** A potential function U on manifold M is a differentiable real-valued function $U : M \to \mathbb{R}$. In other words, a potential function is also a C^1 scalar (potential) field on M. A potential field U is said to be dynamic if U varies with time.

Definition 5.11 ***(Distance vector)*** Given the distance between two points $d(p_0, p_1)$ on connected Riemannian manifold M, by designating p_0 as the *source*, the distance vector from p_0 to p_1 is defined as

$$\mathbf{d}(p_0, p_1) = \log_{p_0}(p_1) \in T_{p_0}M \tag{5.12}$$

where $T_{p_0}M$ denotes the tangent space of M at p_0.

Definition 5.12 ***(Distance between subsets on M)*** Given two subsets $M_0, M_1 \subset M$, the distance between M_0 and M_1 is defined as[3]

$$d(M_0, M_1) = \min_{p_0, p_1} d(p_0, p_1), \forall p_0 \in M_0, p_1 \in M_1 \tag{5.13}$$

Further, By designating M_0 as the *source*, the distance vector $\mathbf{d}(M_0, M_1)$ can be uniquely obtained from $\mathbf{d}(p_0, p_1)$.

Remark 5.2 Since any point set $\{p\} \subset M$ is a subset of M, the previous generalized definitions can be used to measure the distance between a subset M_0 and a point p, denoted as $d(M_0, p)$ with a slight abuse of notation.

Next, we generalize the concept of probability density functions to potential functions such that states, measurements, and constraints are represented geometrically.

---

3 Or $d(M_0, M_1) = \lim \min_{p_0, p_1} d(p_0, p_1), \forall p_0 \in M_0, p_1 \in M_1$, if the minimum does not exist.

A state at a particular time is denoted as a point in the state space. To bring in process noise during state evolution, a random variable is usually used to describe the confidence level of the estimated state, where the mean and covariance are two critical characteristics in Gaussian-distributed states. Probability density functions, which can be represented in DPF form, are required to describe non-Gaussian or multimodal states. Similarly, by considering the measurement noise, measurements are modeled as random variables in measurement space. Thus, these two concepts can be inherently migrated to the DPF formulation with clear probabilistic implications.

Definition 5.13 ***(State (measurement)-sourced DPF)*** Given a point p_0 in m-dimensional state space M (measurement space N), the state (measurement)-sourced DPF is defined for $p \in M$ ($p \in N$) as

$$U(p) = \alpha_{\text{s/m}} K_{\text{s/m}}(\mathbf{d}(p_0, p)) \tag{5.14}$$

where $\alpha_{\text{s/m}}$ is the scaling factor; $K_{\text{s/m}} : \mathbb{R}^m \to \mathbb{R}$ is the kernel function controlling the DPF shape. We call state $p_0 \in M$ (measurement $p_0 \in N$) the *source* of $U(p)$.

Unlike states and measurements, it is impracticable to assign each constraint a probability distribution. It is noted that state (measurement) constraints essentially specify a subset $M_0 \subset M$ ($N_0 \subset N$) in the state (measurement) space, where the distance $d(M_0, p)$ (or $d(N_0, p)$) is well-defined for any non-empty M_0 and $p \in M$ (N_0 and $p \in N$). To represent constraints geometrically, we have the following definition:

Definition 5.14 ***(Constraint-sourced DPF)*** Given a subset M_0 of m-dimensional state space M (N_0 of m-dimensional measurement space N), the state (measurement) constraint-sourced DPF is defined for $p \in M$ ($p \in N$) as

$$U(p) = \alpha_c K_c(\mathbf{d}(M_0, p)) \text{ or } U(p) = \alpha_c K_c(\mathbf{d}(N_0, p)) \tag{5.15}$$

where α_c is the scaling factor; $K_c : \mathbb{R}^m \to \mathbb{R}$ is the kernel function controlling the DPF shape. We call state constraint M_0 (measurement constraint N_0) the *source* of $U(p)$.

5.2.3 DPF-Based Information Fusion

With the DPF representation, we introduce geometric pose estimation, which incorporates states, measurements, and constraints. The procedure is shown in Algorithm 3, for which the main functions are presented in this subsection.

Function `DPF_generation` Given the source and characteristics set, the DPF representing states, measurements, and constraints can be generated from Eqs. (5.14)

Algorithm 3 Geometric Pose Estimation (Geo-PE)

Input: State space M, measurement space N, initial state $\tilde{\mathbf{x}}_0$, measurement $\mathbf{z}_k$, state constraints $M_k \subset M$, measurement constraints $N_k \subset N$, characteristics set chr.

Output: Expectation of the estimated state $\hat{\mathbf{x}}_k$, updated characteristics set chr.

$k = 1$
$U_0^{\mathrm{es}} = \texttt{DPF\_generation}(\tilde{\mathbf{x}}_0, chr)$
while `iteration_condition` **do**
  $U_k^{\mathrm{ps}} = \texttt{DPF\_evolution}(U_{k-1}^{\mathrm{es}})$ // predicted state
  $U_k^{\mathrm{m}} = \texttt{DPF\_generation}(\mathbf{z}_k, chr)$ // measurement
  $U_k^{\mathrm{sc}} = \texttt{DPF\_generation}(M_k, chr)$ // state constraint
  $U_k^{\mathrm{mc}} = \texttt{DPF\_generation}(N_k, chr)$ // measurement constraint
  $U_k^{\mathrm{es}} = \texttt{DPF\_fusion}(U_k^{\mathrm{ps}}, U_k^{\mathrm{m}}, U_k^{\mathrm{sc}}, U_k^{\mathrm{mc}}, chr)$ // estimated state
  $\{\hat{\mathbf{x}}_k, chr\} = \texttt{pose\_recovery}(U_k^{\mathrm{es}}, chr)$
  $k{+}{+}$
end while

and (5.15). For states and measurements, the kernel can be selected to represent probability distributions, among which the Gaussian

$$K_{\mathrm{s/m}}(\mathbf{d}) = \exp\left(-\frac{1}{2}\mathbf{d}^{\mathrm{T}}\boldsymbol{\Sigma}^{-1}\mathbf{d}\right) \tag{5.16}$$

has been extensively used.

For constraints, since a vector field ∇U can be generated from the DPF U using the derivative operator ∇, we design the kernel such that all points are projected toward constraints along geodesics

$$K_{\mathrm{c}}(\mathbf{d}) = -\frac{1}{2}\mathbf{d}^{\mathrm{T}}\mathbf{d} \tag{5.17}$$

which leads to the vector field[4] $\nabla U = -\alpha_{\mathrm{c}}\mathbf{d}$. We further require $0 < \alpha_{\mathrm{c}} < 1$, indicating the strictness of soft constraints. Naturally, two mappings emerge for any point in state space and measurement space, representing state and measurement constraints, respectively

$$\phi^{\mathrm{sc}}(p) = \exp_p(\nabla U^{\mathrm{sc}}), \forall p \in M \tag{5.18}$$

$$\phi^{\mathrm{mc}}(p) = \exp_p(\nabla U^{\mathrm{mc}}), \forall p \in N \tag{5.19}$$

where $\exp_p$ denotes the exponential map at p. As an example, ϕ^{sc} is geometrically illustrated in Figure 5.2.

4 As in Definition 5.14, **d** belongs to the tangent space of a source point of the DPF. However, **d** can be parallelly transported along the geodesic. The parallel transport notation is omitted here.

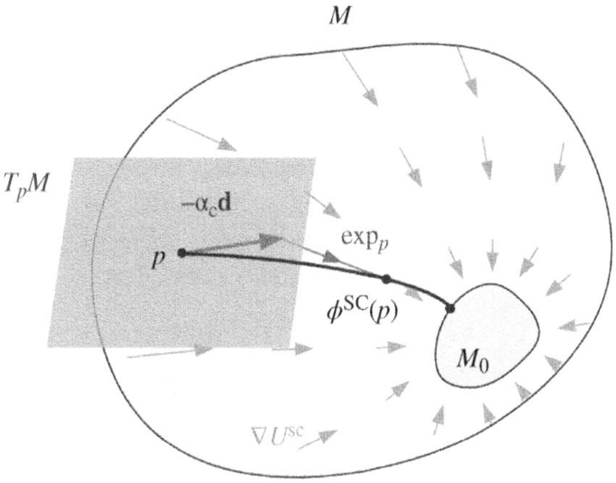

Figure 5.2 Illustration of ϕ^{sc}. A dynamic potential field U^{sc} can be obtained from the distance between state constraints M_0 and any point p on M. Given U^{sc}, we first generate a vector field ∇U^{sc}; then $\phi^{sc}(p)$ is defined as the exponential map of ∇U^{sc} at p.

***Function* `DPF_evolution`** Given the last moment state, state equation (5.9) predicts the current state. By considering $\mathbf{f}_\mathbf{w}$ as constant at each discrete time, given the DPF U^{es}_{k-1} on M, and letting $\phi^{evo} = \mathbf{f}_\mathbf{w} \circ \mathbf{f}$, the DPF evolution for $p \in M$ is defined as

$$U^{es}_k(p) = U^{es}_{k-1}\left(\{\phi^{evo}\}^{-1}(p)\right) \tag{5.20}$$

where the existence of inverse mapping $\{\phi^{evo}\}^{-1}$ is ensured owing to Proposition 5.2.

By writing $\phi^{evo}_* : U^{es}_{k-1} \mapsto U^{es}_k$, it can be derived that

Proposition 5.3 *ϕ^{evo}_* is linear. In other words,*

$$\phi^{evo}_*\left(\alpha_1 U^{es1} + \alpha_2 U^{es2}\right) = \alpha_1\phi^{evo}_*\left(U^{es1}\right) + \alpha_2\phi^{evo}_*\left(U^{es2}\right)$$
$$\text{for any DPF } U^{es1}, U^{es2} \text{ and } \alpha_1, \alpha_2 \in \mathbb{R} \tag{5.21}$$

Proof: For any $p \in M$, $\phi^{evo}_*\left(\alpha_1 U^{es1} + \alpha_2 U^{es2}\right)(p)$

$$= \left(\alpha_1 U^{es1} + \alpha_2 U^{es2}\right)\left((\mathbf{f}_\mathbf{w} \circ \mathbf{f})^{-1}(p)\right) \tag{5.22}$$

$$= \alpha_1 U^{es1}\left((\mathbf{f}_\mathbf{w} \circ \mathbf{f})^{-1}(p)\right) + \alpha_2 U^{es2}\left((\mathbf{f}_\mathbf{w} \circ \mathbf{f})^{-1}(p)\right) \tag{5.23}$$

$$= \alpha_1\phi^{evo}_*\left(U^{es1}\right)(p) + \alpha_2\phi^{evo}_*\left(U^{es2}\right)(p) \tag{5.24}$$

□

Proposition 5.4 *For any DPF U^{es1}, U^{es2}:*

$$\phi^{evo}_*\left(U^{es1}U^{es2}\right) = \phi^{evo}_*\left(U^{es1}\right)\phi^{evo}_*\left(U^{es2}\right) \tag{5.25}$$

Proof: It is straightforward by following the proof for Proposition 5.3. □

Remark 5.3 Mapping ϕ^{evo} is between points on the manifold, while the induced mapping ϕ_*^{evo} is between DPFs on the manifold. Actually, ϕ_*^{evo} is called the *push forward mapping*.

***Function* `DPF_fusion`** State-sourced, measurement-sourced, and constraint-sourced DPFs are fused to generate a new DPF. For any $p \in M$ and $\mathbf{h} : M \to N$, given ϕ^{sc}, ϕ^{mc}, the update rules $\phi_*^{\text{mc}} : U_k^{\text{m}} \mapsto \tilde{U}_k^{\text{m}}$, $\phi_*^{\text{Bayes}} : \tilde{U}_k^{\text{m}} \times U_k^{\text{ps}} \mapsto \tilde{U}_k^{\text{es}}$, and $\phi_*^{\text{sc}} : \tilde{U}_k^{\text{es}} \mapsto U_k^{\text{es}}$ are sequentially defined as

$$\tilde{U}_k^{\text{m}}(\mathbf{h}(p)) = U_k^{\text{m}}\left(\{\phi^{\text{mc}}\}^{-1}(\mathbf{h}(p))\right) \tag{5.26}$$

$$\tilde{U}_k^{\text{es}}(p) = \beta \tilde{U}_k^{\text{m}}(\mathbf{h}(p))\, U_k^{\text{ps}}(p) \tag{5.27}$$

$$U_k^{\text{es}}(p) = \tilde{U}_k^{\text{es}}\left(\{\phi^{\text{sc}}\}^{-1}(p)\right) \tag{5.28}$$

where Eqs. (5.26) and (5.28) aim to incorporate constraints into the measurement- and state-sourced DPFs; Eq. (5.27) follows Bayes' theorem; and β is the normalizing constant.

Remark 5.4 The inverse mappings $\{\phi^{\text{sc}}\}^{-1}$ and $\{\phi^{\text{mc}}\}^{-1}$ exist if an appropriate *normal neighborhood* is selected [Hawking and Ellis, 1973]. In this work, we consider M and N in SE(3) and SO(3) such that the exponential map is well-defined, one-to-one, and onto [Blanco, 2010], which ensures the existence of $\{\phi^{\text{sc}}\}^{-1}$ and $\{\phi^{\text{mc}}\}^{-1}$.

Proposition 5.5 *ϕ_*^{mc}, ϕ_*^{Bayes}, ϕ_*^{sc} are (bi)linear.*

Proof: By omitting the time index k and letting $p' = \mathbf{h}(p)$, for any $p \in M$, we can prove that ϕ_*^{mc} and ϕ_*^{sc} are linear by following the proof for Proposition 5.3. As for ϕ_*^{Bayes}, we first take $\tilde{U}^{\text{m}}$ as constant:

$$\begin{aligned}
&\phi_*^{\text{Bayes}}\left(\alpha_1 U^{\text{ps1}} + \alpha_2 U^{\text{ps2}}\right)(p) \\
&= \beta \tilde{U}^{\text{m}}\left(p'\right)\left(\alpha_1 U^{\text{ps1}}(p) + \alpha_2 U^{\text{ps2}}(p)\right) && (5.29) \\
&= \alpha_1 \beta \tilde{U}^{\text{m}}\left(p'\right) U^{\text{ps1}}(p) + \alpha_2 \beta \tilde{U}^{\text{m}}\left(p'\right) U^{\text{ps2}}(p) && (5.30) \\
&= \alpha_1 \phi_*^{\text{Bayes}}\left(U^{\text{ps1}}(p)\right) + \alpha_2 \phi_*^{\text{Bayes}}\left(U^{\text{ps1}}(p)\right) && (5.31)
\end{aligned}$$

The proof still holds if we take U^{ps} as constant. Thus ϕ_*^{Bayes} is bilinear. □

***Function* `pose_recovery`** It serves as the inverse problem of `DPF_generation` to convert the DPF back into statistical characteristics. Given a DPF, the variance, expectation, and covariance matrix are well-defined, as in Section 5.1. However, unlike parameter estimation in Euclidean space, the fused DPF may not be in good properties due to the nontrivial topological structure of the background manifold. Thus, analytic results are usually non-feasible, and numerical methods need to be developed.

5.2.4 Approximation of Geometric Pose Estimation

We have proposed a DPF-based information fusion scheme, but two issues remain: (i) explicit analytical forms are usually hard to find for DPFs; and (ii) the noise items $\mathbf{f_w}$ and $\mathbf{h_v}$ are practically unknown for a particular moment, but their statistical characteristics can be modeled. Since Geo-PE is based on mappings between DPFs, we now discuss how to approximate such mappings so that Geo-PE would be adaptive to various state and output equations.

By comparing Definitions 5.10 and 5.2, for any $p \in M$, it is obvious that a DPF can be linearly mapped to a PDF up to a scaling factor. Thus, any DPF can be represented as a set of samples generated from a random sampling scheme, where the probability density of samples approximates the DPF as the sample number goes to infinity. As mentioned in Remark 5.3, a push forward mapping ϕ_* between DPFs can be inherently induced from a homeomorphism ϕ between manifolds. With the relation between ϕ_* and ϕ, samples together with their weights $\{p_i, w_i\}$ are mapped to $\{\phi(p_i), w_i\}$ to approximate the DPF mapping ϕ_*. Kernel density estimation (KDE) ensures the convergent recovery of the DPF as the sample number goes to infinity [Ma and Fu, 2011]. The approximation process is illustrated in Figure 5.3.

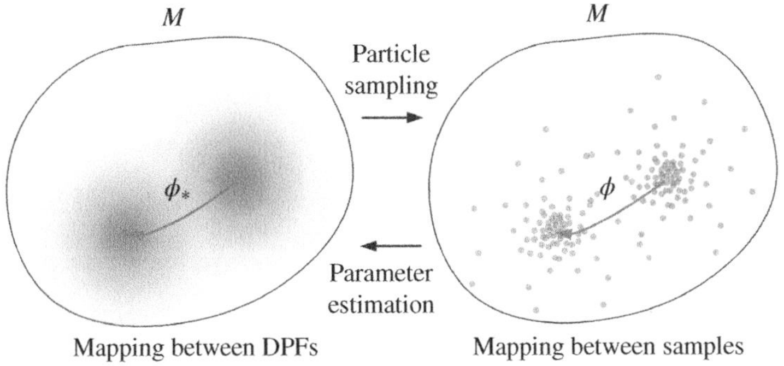

Figure 5.3 Approximating mappings between DPFs with mappings between samples. The gradient arrow on the left represents the induced push forward mapping ϕ_*, while the right gradient arrow indicates the corresponding mapping between samples in order to approximate ϕ_*.

Remark 5.5 Essentially, benefiting from Propositions 5.3–5.5, the approximation first decomposes the DPF with multiple distributions, then applies the fusion scheme for each decomposed distribution, and finally composes them again for the overall distribution.

Given this discussion and inspired by the Monte Carlo method, the following approximation scheme is proposed:

Particle Sampling A set of equally weighted particles $\{(p_i, w_i)\}$ is generated according to *chr* and $\tilde{\mathbf{x}}_0$ to represent U_0^{es}.

Particle Evolution For each k, the particle is updated as

$$(p_i, w_i) \leftarrow \left(\phi^{\mathrm{evo}}(p_i), \frac{w_i}{\sum_i w_i}\right) \tag{5.32}$$

to represent U_k^{es}.

Particle Weight Update To incorporate measurements, the weight of each particle is updated as

$$(p_i, w_i) \leftarrow \left(p_i, w_i \tilde{U}_k^{\mathrm{m}}\left(\mathbf{h}(p_i)\right)\right) \tag{5.33}$$

with $\tilde{U}_k^{\mathrm{m}}$ from Eq. (5.26).

Particle Projection The state constraints are incorporated according to the updating rule

$$(p_i, w_i) \leftarrow \left(\phi^{\mathrm{sc}}(p_i), w_i\right) \tag{5.34}$$

Parameter Estimation Given $\{(p_i, w_i)\}$, a necessary condition of the expectation is[5]

$$\hat{p} \in \mathbb{E}[\{(p_i, w_i)\}] \Rightarrow \sum_i w_i \mathbf{d}(\hat{p}, p_i) = 0 \tag{5.35}$$

and an iterative algorithm has been obtained in Pennec (2004) to obtain the second-order approximation of the expectation:

$$\hat{p} \leftarrow \exp_{\hat{p}}\left(\frac{\sum_i w_i \mathbf{d}(\hat{p}, p_i)}{\sum_i w_i}\right) \tag{5.36}$$

The estimated pose is then represented as $\hat{\mathbf{x}}_k = \hat{p}$ in a coordinate system. The covariance matrix can be computed from the discrete version of Eq. (5.8). Finally, let $k = k + 1$, and go to *Particle Evolution* for the next iteration.

---

5 under the assumption that the variance is finite everywhere and the cut locus has a null probability measure.

5.3 VO-Heading-Map Pose Estimation for Ground Vehicles

Herein, VO measurements, heading measurements from the attitude and heading reference system (AHRS), and road constraints are combined to provide pose estimations for ground vehicles. Fortunately, the left-invariant metric on the Lie group enables single-chart coverage for the whole group.[6] Thus, exponential and logarithmic maps between Lie group and Lie algebra are used for a simplified implementation.

5.3.1 System Modeling

We represent the state as $g = (\mathbf{t}, \mathrm{R}) \in SE(3)$, where $\mathbf{t} \in \mathbb{R}^3$, $\mathrm{R} \in SO(3)$ are the translation vector and rotation matrix, respectively. The group operation is

$$g_1 g_2 = \left(\mathbf{t}_1 + \mathrm{R}_1 \mathbf{t}_2, \mathrm{R}_1 \mathrm{R}_2\right) \tag{5.37}$$

where $g_1 = \left(\mathbf{t}_1, \mathrm{R}_1\right)$, $g_2 = \left(\mathbf{t}_2, \mathrm{R}_2\right)$, and the inverse element is $g^{-1} = \left(-\mathrm{R}^\top \mathbf{t}, \mathrm{R}^\top\right)$. The state evolution is modeled as

$$g_{k+1} = g_k g_{k,k+1} \tag{5.38}$$

where g_k denotes the state at time k; $g_{k,k+1}$ represents the VO pose transformation from k to $k+1$.

To allow $g_{k,k+1}$ randomness, we generate $g_{k,k+1}$ from

$$g_{k,k+1} = \exp\left(\sum_{i=1}^{6} x_i \mathbf{E}_i\right) \tag{5.39}$$

where $[x_1, \cdots, x_6]^\top$ is a Gaussian-distributed random vector with covariance matrix $\boldsymbol{\Sigma}_{\mathrm{VO}}$, and

$$\mathbf{E}_1 = \begin{bmatrix} 0 & 0 & 0 & 0 \\ 0 & 0 & -1 & 0 \\ 0 & 1 & 0 & 0 \\ 0 & 0 & 0 & 0 \end{bmatrix}, \mathbf{E}_2 = \begin{bmatrix} 0 & 0 & 1 & 0 \\ 0 & 0 & 0 & 0 \\ -1 & 0 & 0 & 0 \\ 0 & 0 & 0 & 0 \end{bmatrix}, \mathbf{E}_3 = \begin{bmatrix} 0 & -1 & 0 & 0 \\ 1 & 0 & 0 & 0 \\ 0 & 0 & 0 & 0 \\ 0 & 0 & 0 & 0 \end{bmatrix}$$

$$\mathbf{E}_4 = \begin{bmatrix} 0 & 0 & 0 & 1 \\ 0 & 0 & 0 & 0 \\ 0 & 0 & 0 & 0 \\ 0 & 0 & 0 & 0 \end{bmatrix}, \mathbf{E}_5 = \begin{bmatrix} 0 & 0 & 0 & 0 \\ 0 & 0 & 0 & 1 \\ 0 & 0 & 0 & 0 \\ 0 & 0 & 0 & 0 \end{bmatrix}, \mathbf{E}_6 = \begin{bmatrix} 0 & 0 & 0 & 0 \\ 0 & 0 & 0 & 0 \\ 0 & 0 & 0 & 1 \\ 0 & 0 & 0 & 0 \end{bmatrix}.$$

Since the only measurement in this work is the rotation from the AHRS, the ideal output equation can be written as mapping $(\mathbf{t}, \mathrm{R}) \mapsto \mathrm{R}$. To obtain particle

6 Except for some singularities.

weights, a Gaussian distributed DPF on SE(3) is generated from [Wang and Chirikjian, 2006, Wang and Chirikjian, 2008]

$$\mathbb{P}(g|\mu, \boldsymbol{\Sigma}_{\mathrm{AHRS}}) = \alpha_{\mathrm{m}} \exp\left\{ -\frac{1}{2}\left[\left(\log(\mu^{-1}g)\right)^{\vee}\right]^{\mathsf{T}} \boldsymbol{\Sigma}_{\mathrm{AHRS}}^{-1} \left[\left(\log(\mu^{-1}g)\right)^{\vee}\right] \right\} \tag{5.40}$$

for $g, \mu \in \mathrm{SE}(3)$, where α_{m} is the normalizing factor such that $\int_G f(g|\mu, \boldsymbol{\Sigma}_{\mathrm{AHRS}})dg = 1$; $\boldsymbol{\Sigma}_{\mathrm{AHRS}}$ is the covariance matrix indicating AHRS measurement noise; dg denotes the integration measure (volume element) for SE(3); and the notation "$^{\vee}$" represents the Lie algebra as an isomorphic vector [Barfoot and Furgale, 2014]. To avoid integration, α_{m} can be approximated as $\alpha_{\mathrm{m}} = (8\pi^3 |\det \boldsymbol{\Sigma}_{\mathrm{AHRS}}|^{\frac{1}{2}})^{-1}$ by assuming the distribution is tightly concentrated on SE(3).

5.3.2 Road Constraints

Maps containing road constraints are downloaded from OSM for each testing sequence. As a precise 3D map with altitude is currently unavailable, road constraints are represented in the East-North plane, as in Cui and Ge (2003) and Jiang et al. (2018). To accelerate pose estimation, a unique vector field for each map needs to be generated before the experiments.

First, the metric OSM is converted to a binary image indicating drivable roads.[7] A distance transform [Breu et al., 1995] is applied to the binary image so that a discrete distance potential field that represents the distance to the nearest road is generated for each pixel. Then, a negative gradient vector can be obtained for each

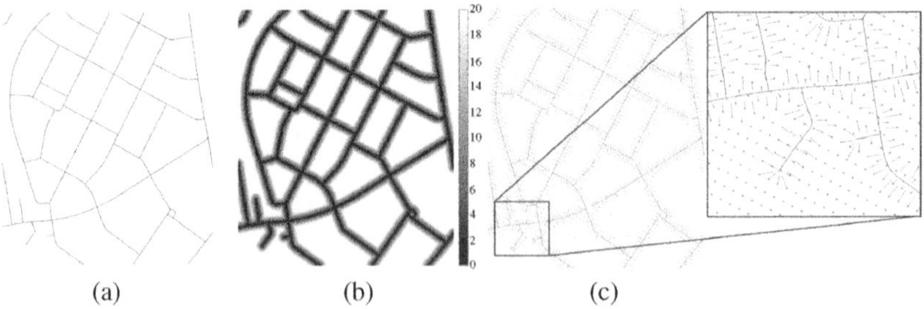

Figure 5.4 Generating road constraints for KITTI sequence 00 from OSM. (a) Parsed OSM displaying drivable roads only. (b) Distance potential field after distance transformation. Numbers in the color bar denote the distance to the nearest road. The maximum distance threshold is set to 20 pixels. (c) The vector field (down-sampled for clearness) is used for particle projection. The vector field is illustrated using dark gray line segments dotted at each pixel.

7 OSM is parsed with the code from https://github.com/johnyf/openstreetmap.

pixel from the distance potential field. Based on the vector field corresponding to the road map, particles are projected toward roads at a discount rate $0 \leq \alpha_c \leq 1$ such that the road constraints are incorporated. The generation of road constraints is shown in Figure 5.4.

Remark 5.6 The discount rate α_c stands for the softness of road constraints. Road constraints will be noneffective if $\alpha_c = 0$, while $\alpha_c = 1$ brings hard constraints to pose estimation and will cause $\{\phi^{sc}\}^{-1}$ to be undefined.

5.3.3 Parameter Estimation on SE(3)

Based on Eq. (5.36) and the left-invariant metric, given a particle set $\{g_i, w_i\}$, by selecting any initial $\hat{g} \in \{g_i\}$, the expectation $\hat{g}$ in the Lie group can be estimated iteratively as

$$\hat{g} \leftarrow \hat{g} \exp\left(\frac{\sum_i w_i \log(\hat{g}^{-1} g_i)}{\sum_i w_i}\right) \tag{5.41}$$

5.4 Experiments on KITTI Sequences

The data comes from the public KITTI dataset [Geiger et al., 2013], where SVO results are obtained from `libviso2` [Geiger et al., 2011a], and AHRS measurements are simulated from the ground truth with manually added Gaussian noises $\mathcal{N}(\mathbf{0}, \text{diag}(\sigma_r^2, \sigma_p^2, \sigma_y^2))$ for roll, pitch, and yaw. In this section, we first show the overall performance of the proposed approach; then the influence of heading measurement error $\text{diag}(\sigma_r^2, \sigma_p^2, \sigma_y^2))$ is presented. Finally, the parameter settings including number of particles n_p, covariance matrices $\boldsymbol{\Sigma}_{VO}$ and $\boldsymbol{\Sigma}_{AHRS}$, discount rate α_c, and road map resolution are discussed separately.

5.4.1 Overall Performance

Results of the multi-sensor Geo-PE are shown in Figure 5.5 and Figure 5.6 and listed in Table 5.2. As Geo-PE takes VO as an integrated sensor instead of considering image features, we select another "loose-coupled" framework, loosely coupled heading reference pose estimation (LC-HRPE) from Chapter 2, as a benchmark. The Euler angle noises are set to $\sigma_r = \sigma_p = \sigma_y = 5°$. At each iteration, 100 particles are generated. As we have measurements for rotation only, we set $\boldsymbol{\Sigma}_{VO} = 10^{-4}\text{diag}(1, 1, 1, 0, 0, 0)$ and $\boldsymbol{\Sigma}_{AHRS} = 0.01\text{diag}(1, 1, 1, 0, 0, 0)$ to avoid wasting particles. The discount rate is $\alpha_c = 0.05$.

Regarding quantitative results, Geo-PE without a map performs slightly better than LC-HRPE for both translational and rotational errors, except for

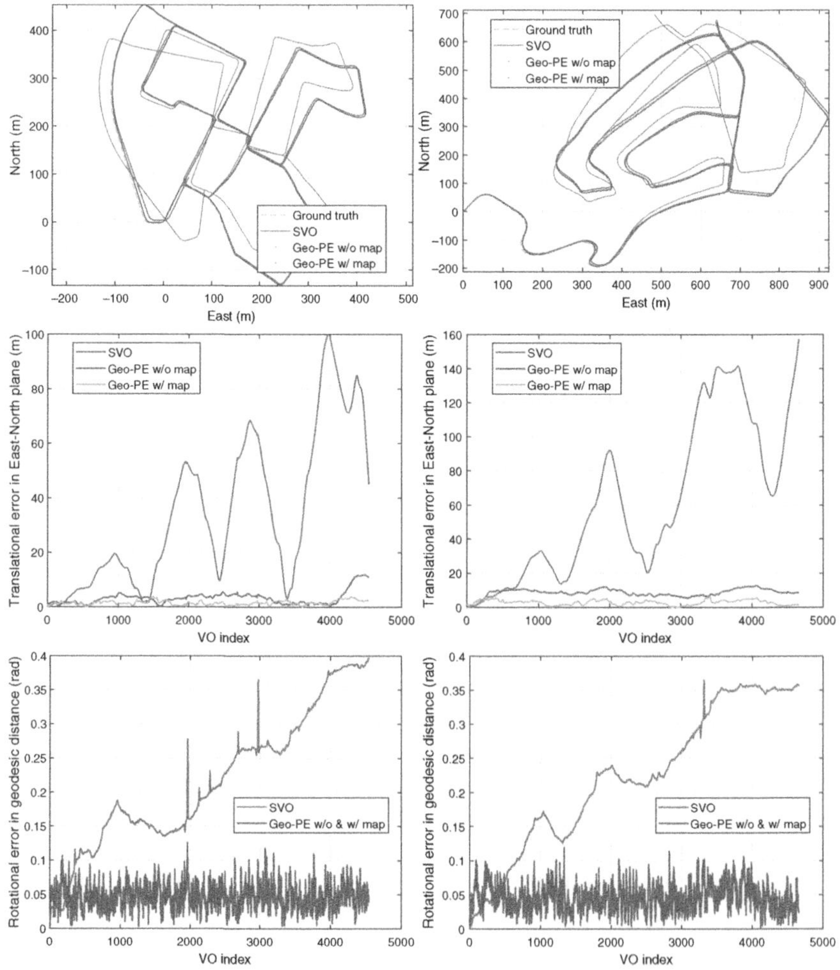

Figure 5.5 Trajectories (upper), translational error (middle), and rotational error (bottom) of SVO, Geo-PE without a map, and Geo-PE with a map for KITTI sequences 00 (left) and 02 (right).

the translational error of KITTI sequence 02. Based on the results of Geo-PE with a map, the road constraints significantly improve translational estimation performance. Note that road constraints project only the translational part of the pose to drivable roads, and this means rotational errors with and without the map would be identical.

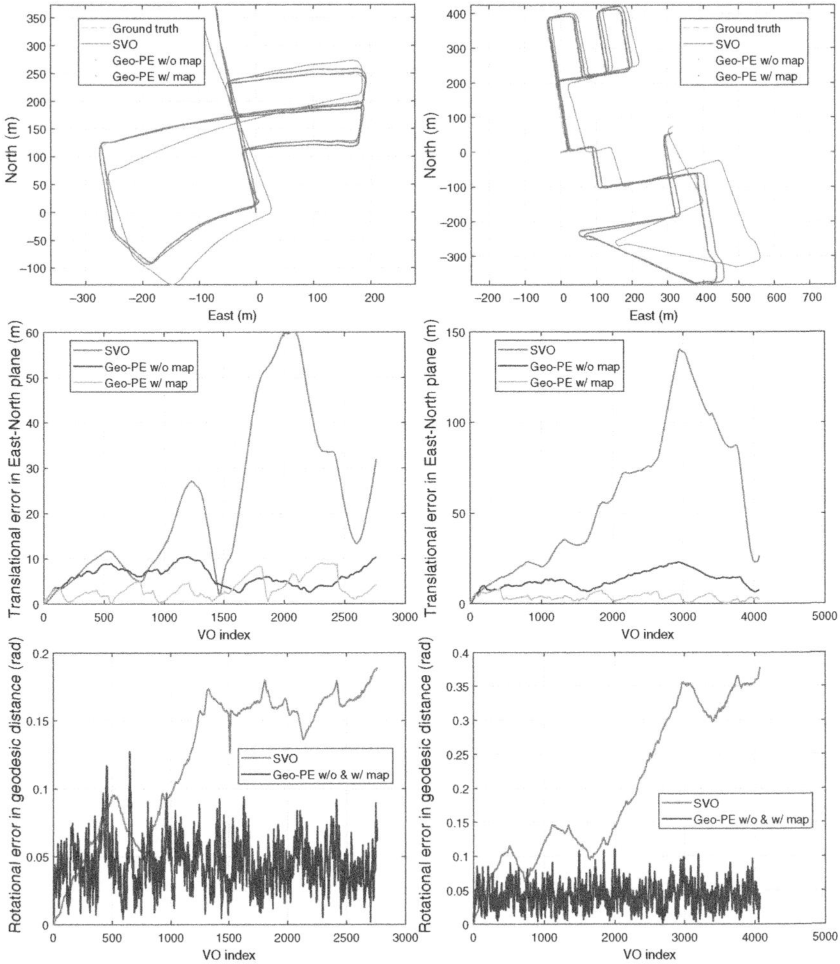

Figure 5.6 Trajectories (upper), translational error (middle), and rotational error (bottom) of SVO, Geo-PE without a map, and Geo-PE with a map for KITTI sequences 05 (left) and 08 (right).

Regarding qualitative performance, with the help of AHRS measurements, rotation drift can be corrected in Geo-PE, but the random error still exists. As a benefit of eliminating rotation drift, the translation error has been greatly reduced. It is observed that sometimes Geo-PE with a map brings more substantial translational error because the vehicle may not be close to the center of the road, which is usually modeled in a road map.

Table 5.2 Pose estimation results of the proposed approach and LC-HRPE on KITTI sequences 00, 02, 05, and 08. Rotational errors of Geo-PE without a map and Geo-PE with a map are identical and have been replaced with "-" for Geo-PE without a map.

Seq	Dist (m)	Map size: Metric(m)/ Grid(px)	Geo-PE w/o map: Trans(m)/Yaw(deg) Avg error	Geo-PE w/o map: Trans(m)/Yaw(deg) Std dev	Geo-PE w/ map: Trans(m)/Yaw(deg) Avg error	Geo-PE w/ map: Trans(m)/Yaw(deg) Std dev	LC-HRPE: Trans(m)/Yaw(deg) Avg error	LC-HRPE: Trans(m)/Yaw(deg) Std dev
00	3,724	569.9 × 656.8 /673 × 776	3.24/-	2.41/-	1.52/1.30	0.79/0.93	10.39/1.49	7.65/1.12
02	5,067	1000.8 × 946.3 /1182 × 1118	8.58/-	2.13/-	2.82/1.30	1.39/0.94	4.10/1.46	1.72/1.14
05	2,205	518.0 × 489.8 /777 × 735	6.05/-	2.18/-	3.48/1.22	2.35/0.87	19.61/1.61	10.34/1.25
08	3,222	526.0 × 879.4 /505 × 844	13.02/-	4.65/-	3.56/1.16	1.76/0.90	23.75/1.64	12.81/1.26

5.4.2 Influence of Heading Error

Figure 5.7 shows translational and rotational errors with different manually added Gaussian Euler angle noises. With the help of road constraints, the average translational error is less than 5 m with heading measurements under Euler angle noise $\sigma_r = \sigma_p = \sigma_y = 15°$. We evaluate the rotational error based on the geodesic distance (length) such that errors in roll, pitch, and yaw are all considered based on a unified criterion. As predicted, the rotational error grows with increasing heading measurement error. Note that the error belongs to random error, which excludes rotation drift; thus, estimated trajectories are well-aligned to the global map compared to the SVO approach.

5.4.3 Influence of Road Map Resolution

The road map resolution is defined as the ratio of the metric OSM length (width) to the binary image length (width) with the units m/pixel. Figure 5.8 shows the translational errors with different road map resolutions; no difference is seen for rotational errors. It is obvious that a fine map results in smaller errors and standard deviation in translation estimation than a coarse map. However, a fine map also consumes a large amount of memory, which may limit its implementation, especially for embedded systems.

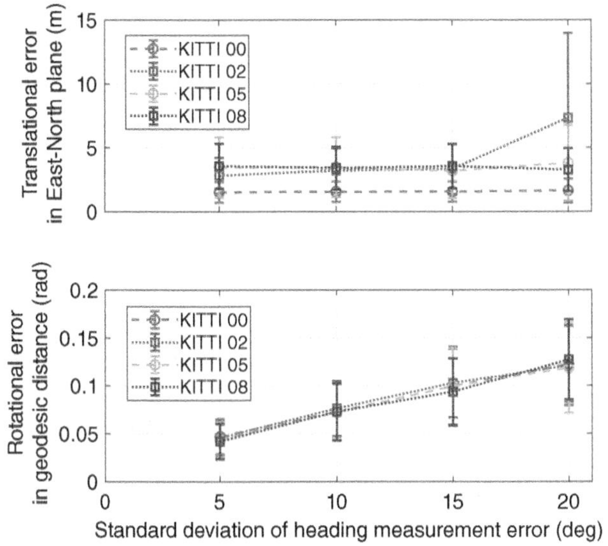

Figure 5.7 Estimation accuracy with respect to AHRS measurement error.

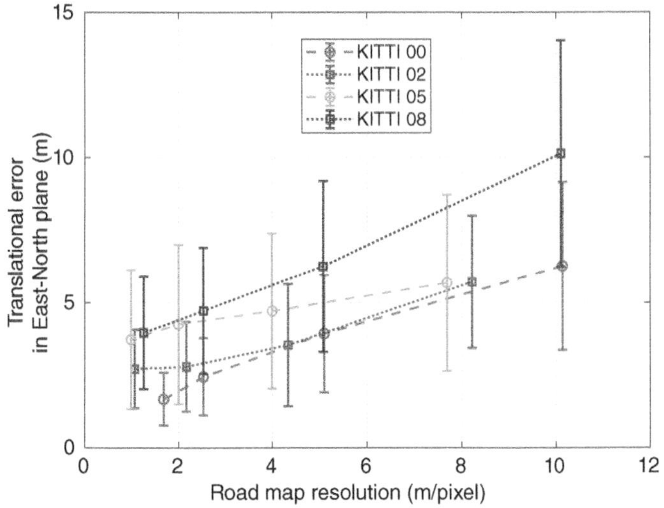

Figure 5.8 Translational estimation accuracy with respect to road map resolution.

5.4.4 Influences of Parameters

Figure 5.9 shows translational and rotational errors with respect to the number of particles n_{p}. Thanks to the road constraints, the translational error remains roughly unchanged. As n_{p} increases, the rotational error decreases with a smaller standard deviation, but more particles lead to lower computing efficiency. For example, with MATLAB R2017b, the average processing rate for Geo-PE with a map (excluding VO) drops from 70.4 Hz (25 particles) to 10.2 Hz (200 particles) in KITTI sequence 00 on a mobile workstation with an Intel Core i5-8250U CPU and 8 GB RAM.

Covariance matrices $\boldsymbol{\Sigma}_{\text{VO}}$ and $\boldsymbol{\Sigma}_{\text{AHRS}}$ reflect the degree of confidence in VO and AHRS measurement, respectively. Both parameters influence approximated probability distributions, but in distinct ways: by adjusting $\boldsymbol{\Sigma}_{\text{VO}}$, the particle dispersity is different, but identical weights are assigned for each particle, while different $\boldsymbol{\Sigma}_{\text{AHRS}}$ leads to invariant particle poses with redistributed weights. A large $\boldsymbol{\Sigma}_{\text{VO}}$ is unnecessary because the transformation between frames should be with little error, and more particles are needed to describe a spread distribution. However, a small $\boldsymbol{\Sigma}_{\text{VO}}$ may fail the AHRS assistance as few particles will fall into the neighborhood of the AHRS measurements. Meanwhile, a large $\boldsymbol{\Sigma}_{\text{AHRS}}$ will invalidate the tightly concentrated assumption so that the distribution is no longer Gaussian on SE(3). An improper $\boldsymbol{\Sigma}_{\text{AHRS}}$ will cause little difference in the particle weight redistribution, which may weaken the role of AHRS measurements.

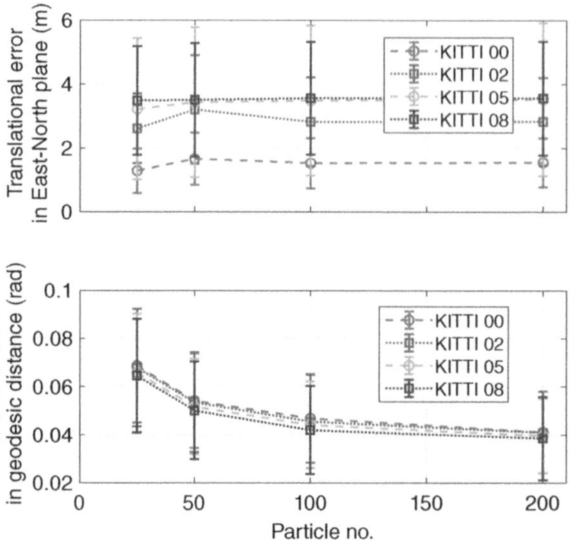

Figure 5.9 Estimation accuracy with respect to the number of particles.

Theoretically, setting the discount rate $\alpha_c = 1$ forces all particles to be located on the road. However, it has been observed that a large α_c generally worsens estimation results for the following reasons. (i) Because road constraints are modeled based on the road center, an inherent translational error is introduced if hard road constraints are applied. (ii) In a dense road network, it is difficult to track the vehicle's position with frequent road changing, since road changing means the behavior of leaving the current road and joining another road, and this "leaving-joining" behavior has not been considered in this work.

5.5 Experiments on the NTU Dataset

To validate the proposed approach, self-collected stereo image sequences in challenging scenarios near Nanyang Technological University (NTU) have been used. The vehicle's position (latitude, longitude, altitude) and heading in the ENU frame have been obtained from the DJI A3 module. All data is synchronized before pose estimation for evaluation. In experiments, we use headings from DJI A3 as measurements, and only translational errors are assessed by comparing the vehicle's GPS positions with estimated positions. Readers may refer to Chapter 2 for details on the experimental platform and the NTU sequences.

Although scale ambiguity does not exist in SVO, errors from feature extraction, matching, and camera calibration may lead to fluctuating scale in translation. Hence, we set $\mathbf{\Sigma}_{\mathrm{VO}} = 0.001\mathrm{diag}(1,1,1,1,1,1)$ and $\mathbf{\Sigma}_{\mathrm{AHRS}} = 0.001\mathrm{diag}(1,1,1,0,0,0)$ to allow uncertainty in both rotational and translational measurements. During the experiments, 200 particles are generated at each iteration. The discount rate is $\alpha_c = 0.04$. The maximum distance threshold of the road constraints is set to 8 pixels.

5.5.1 Overall Performance

Similar to the results in the KITTI dataset, Table 5.3 indicates that Geo-PE without a map performs slightly better or worse than LC-HRPE, while Geo-PE with a map improves translational estimation performance significantly. The robustness of translation estimation in Geo-PE with a map is also increased, as smaller standard deviations are observed for translational errors, especially in NTU sequences 01 and 04. The vehicle trajectories and translational error comparison in Figure 5.10 also show that rotation drift is eliminated with heading measurements and reduced translational error after incorporating road constraints.

5.5.2 Phenomena Observed During Experiments

It is observed in Figure 5.10 that dual roads are modeled as two neighboring parallel lines in OSM of NTU sequences 01 and 02. The position may gradually be

Table 5.3 Pose estimation results of the proposed approach and LC-HRPE on NTU sequences 01-04.

Seq	Dist (m)	Map size: Metric(m)/ Grid(px)	Geo-PE w/o map: Trans(m)		Geo-PE w/ map: Trans(m)		LC-HRPE: Trans(m)	
			Avg error	Std dev	Avg error	Std dev	Avg error	Std dev
01	1,244	1037.2×615.6 /1225×728	27.11	21.61	12.48	4.93	22.3	10.62
02	1,249	1230.8×601.1 /1453×711	16.61	11.87	12.93	7.94	19.53	8.13
03	860	810.2×614.5 /957×726	30.40	8.75	13.02	5.46	21.41	8.19
04	1,204	740.1×645.7 /875×763	34.72	30.79	10.66	5.39	31.96	26.76

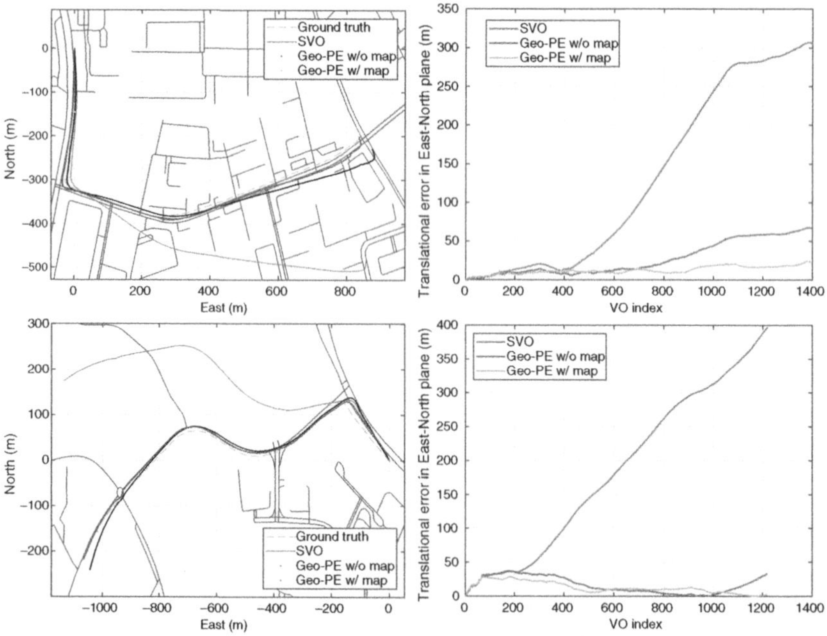

Figure 5.10 Trajectories (left) and translational error (right) of SVO, Geo-PE without a map, and Geo-PE with a map for NTU sequences 01, 02, 03, and 04.

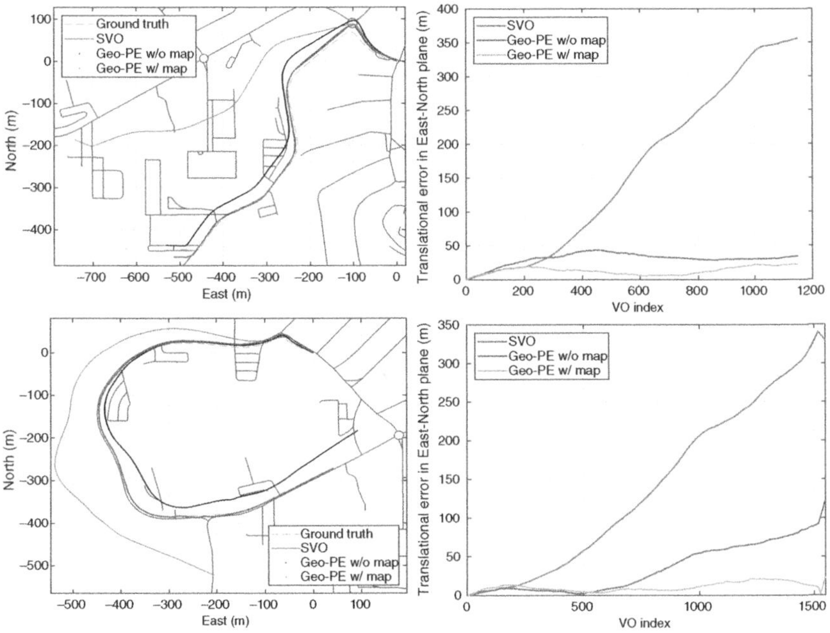

Figure 5.10 (*Continued*)

projected to the opposite lane, as shown in the second half trajectory of NTU 01 when the translational deviation of heading-assisted SVO outstrips the projection distance toward road constraints at each iteration. This error could be corrected if information about the road driving direction was widely provided in OSM. The last turn of Geo-PE with a map in NTU sequence 03 shows that it is very difficult to bound translation drift in the longitudinal direction because there is no such "absolute reference" applied in translation in this work. It is believed that measurements from translation reference sensors (e.g. GPS) would be favorable for further improving pose estimation performance.

5.6 Conclusion

In this chapter, a geometric pose estimation scheme (Geo-PE) has been proposed, where states, measurements, and constraints are consistently represented as DPFs in state space and measurement space. The DPF formulations of state equations, output equations, state constraints, and measurement constraints have been derived such that Bayesian inference can be used for constrained pose estimation. Experiments in many stereo visual odometry sequences have been done to validate the effectiveness of Geo-PE, indicating that rotational drift is eliminated and translational errors are greatly reduced with the help of heading measurements and road maps.

Part II

Secure State Estimation for Mobile Robots

6

Filter-Based Secure Dynamic Pose Estimation

6.1 Introduction

Autonomous vehicles (AVs) are capable of sensing the environment and navigating without human input, which offers the possibility of fundamentally changing transportation by averting deadly crashes, providing critical mobility to the elderly and disabled, increasing road capacity, saving fuel, and lowering emissions [Fagnant and Kockelman, 2015]. Concomitantly, AVs have raised some complex questions, e.g. security, licensing, privacy, legal and insurance regulations, etc. However, safety is the top priority of AVs, and the security of AVs is one of the key challenges that hinder the adoption of such technology since any failure of AVs may result in severe human injuries or even death. Hence, this chapter aims at developing a resilient and efficient run-time estimation algorithm for AVs that provides a performance guarantee in the presence of malicious sensor attacks.

The navigation system plays an essential role in an AV, as it determines "where it is" and plans "where to go." Thus, the navigation system on an AV must be sufficiently robust to prevent a sensor failure in the vehicle. In this chapter, we focus on vehicle pose estimation, which uses data collected from different sensors such that the translational and rotational coordinates can be acquired [Jo et al., 2016]. To accomplish autonomous tasks, AVs should have the capability to determine their pose by using sensors mounted on the vehicle. Accurate pose estimation of AVs remains a significant challenge, especially given the existence of sensor attacks. Although many studies on sensor diagnostics [Loureiro et al., 2014, Hansen and Blanke, 2014, Suarez et al., 2016] and fault-tolerant estimation [Hajiyev et al., 2012, Miranda and Edelmayer, 2012, Liu and Shi, 2013] have been conducted over the past few decades, it is essential to consider pose estimation problems with possible sensor attacks, given the rapid development of the Internet of Vehicles (IoV), where attacks may happen more frequently and seriously.

In accordance with Cardenas et al. (2008), attack types can be arguably classified as deception attacks [Ma et al., 2017a, Liu et al., 2018a, Ding et al., 2018a]

Multimodal Perception and Secure State Estimation for Robotic Mobility Platforms, First Edition.
Xinghua Liu, Rui Jiang, Badong Chen, and Shuzhi Sam Ge.

and denial-of-service (DoS) attacks [Gope et al., 2017], which result in the loss of integrity and availability of sensor-control data, respectively. From the viewpoint of impacts, deception attacks can include an incorrect sensor measurement, an inaccurate timestamp, or the incorrect identity of the sending device. By utilizing the techniques of dynamic programming and Lyapunov theory, some preliminary results have been reported in Amin et al. (2013b) for the case of DoS attacks and Amin et al. (2013a) for deception attacks. Due to protective equipment or special software in the attacked systems, attacks may occur randomly, and their success is largely dependent on the conditions of the cyber-physical systems. In terms of this viewpoint, Amin et al. (2013b) first created a Bernoulli process with known statistical information to govern DoS attacks in a random way. Subsequently, the Bernoulli process was also used to govern randomly occurring deception attacks [Ding et al., 2016].

In this chapter, we assume that the adversary tries to forge the sensory data of AVs by spoofing or physically altering the sensors: i.e. an incorrect sensor measurement, which can be regarded as a deception attack. Thus, possible sensor attacks on AVs can be considered randomly occurring deception attacks.

This chapter focuses on securely estimating the translational and rotational states in a simplified 2D scenario. Distinguished from a conventional Kalman filter, a Kalman-type secure recursive estimator has been designed to provide estimation allowing for possible attacks on sensors. By solving several matrix difference equations, the upper bound of the estimation error covariance is guaranteed and correctly updated during the recursive process.

The highlights of this chapter can be summarized as follows:

1. The system, measurement, and attacks have been modeled such that the secure dynamic pose estimation problem has been formulated for AVs. A study of the problem has practical significance in building a secure navigation system, especially for vehicles connected to the IoV.
2. We introduce a secure pose estimation scheme such that an upper bound of the estimation error covariance is given as a metric for estimator performance or sensor data reliability. Parameter selection for the proposed estimator is discussed for the convenience of practical implementation.
3. In contrast to most of the secure state estimators for cyber-physical systems, the proposed Kalman-type secure recursive estimator can be solved very efficiently in real time, which satisfies the real-time requirement of AVs.
4. Simulated comparisons have been made between the proposed approach and a conventional Kalman filter to demonstrate the excellence of the proposed estimator under attack.

The remainder of this chapter is organized as follows. In Section 6.2, we review the related work in secure dynamic pose estimation. Section 6.3 formulates the problem by modeling the system, measurements, and attacks in preparation for proposing the secure dynamic estimator. The estimator design and mathematical proof are presented in Section 6.4, after which Section 6.5 discusses parameter selection. Experimental validation and results are shown in Section 6.6. Finally, Section 6.7 concludes the chapter.

Notation: The following notations are used throughout this chapter. We use $\mathbb{R}^n$ to denote n-dimensional Euclidean space and $\mathbb{R}^{m \times n}$ for the set of all $m \times n$ matrices. $\mathbb{E}$ denotes the mathematical expectation operator of an underlying probability space, which will be clear from the context. We let I be the identity matrix with the proper dimensions. Let $||\mathbf{x}||$ and $||\mathbf{A}||$ be the Euclidean norm of a vector $\mathbf{x}$ and a matrix $\mathbf{A}$, respectively. The superscripts $^\top$ and $^{-1}$ denote matrix transposition and matrix inverse, respectively.

6.2 Related Work

Cyber-physical systems (CPS), which refer to the embedding of widespread sensing, networking, computation, and control into physical spaces, now play a crucial role in many areas. Significant efforts have been invested in secure dynamic state estimation against sensor attacks in CPS in the past few years, e.g. through graph-theoretic methods [Bi and Zhang, 2014], by exploring the sparsity of the attack signals [Sandberg et al., 2010a, Fawzi et al., 2014a, Liu et al., 2017], and using game-theoretic approaches [Li et al., 2015]. The design of a state estimator against attacks, which can tolerate a small portion of the sensory data being altered, has also been extensively studied in the literature [Huber and Ronchetti, 2009, Bezzo et al., 2014, Lee et al., 2015] and the references therein. However, it must be noted that the problem of designing a secure state estimator for CPS is much more challenging since the bias injected by an adversary can accumulate in the state estimation, and the adversary can potentially exploit this fact to introduce a large or even unbounded estimation error.

As an important application area of CPS, AVs have become the focus of researchers, and some interesting attempts at AVs have been reported in the literature; refer to Li et al. (2014), Han and Ge (2015), and Jiang et al. (2019b) and the references therein. Motivated by the principle of secure state estimation for CPS, we examine the problem of secure pose estimation for AVs in this chapter. To the best of the authors' knowledge, the problem of secure dynamic pose estimation for AVs has not been properly investigated yet, not to mention the design of a robust pose estimator against attacks on AVs. Such a situation has motivated the present investigation. To solve the problem of bias accumulation in the dynamic state estimation for CPS, the combinatorial optimization-based approach is proposed in Fawzi et al. (2014a) and Pasqualetti et al. (2013). These approaches usually iterate over all possible attack scenarios, i.e. the set of compromised sensors and actuators, and decide which is most likely given the sensory data. The main merit of such approaches is that the stability of the estimator is guaranteed: i.e. an upper bound for the estimation error covariance is guaranteed. However, since the number of possible attack scenarios is combinatorial, these approaches often scale poorly and may not be usable for AVs due to real-time constraints.

In this chapter, we consider the impact of randomly occurring deception attacks (possible sensor attacks) on the design of a secure dynamic pose estimator for AVs. By utilizing the Kalman filter algorithm combined with matrix inequality techniques, we propose a Kalman-type secure recursive estimation algorithm, and an upper bound of estimation error covariance is derived by selected optimal estimator parameters. One of the primary benefits of our proposed scheme is that it can be solved very efficiently in real time and is suitable for recursive computation in online applications. Furthermore, under malicious sensor attacks, we characterize the estimation performance of our proposed secure pose estimator by experimental validation.

6.3 Problem Formulation

In this section, we first model the dynamic system and measurements of pose estimation for AVs. The attack model is subsequently introduced to provide the secure dynamic pose estimation problem formulation.

Since the chapter mainly discusses improvement in the algorithm, we model a typical vehicle with sensors that are frequently used in AV pose estimation. The proposed method is general and can be applied to other sensors or vehicles as long as the system model, measurement model, and attack model can be explicitly derived using the format in the chapter under prescribed assumptions.

6.3.1 System Model

We confine the reference frames used in this chapter to the global frame, body frame, and sensor frames. The global frame does not move with the vehicle, but the body frame and sensor frames do. Because the sensors are all mounted on the vehicle, the transformation matrix between the sensor frames and body frame is time-invariant, and the transformation matrix is usually determined from calibration prior to vehicle operation. Thus, we only consider the global frame and body frame in the remaining part of the chapter for convenient representation.

As demonstrated in Figure 6.1, under the low dynamics assumption and the principal motion assumption [Kelly, 2013], and by assuming that the vehicle

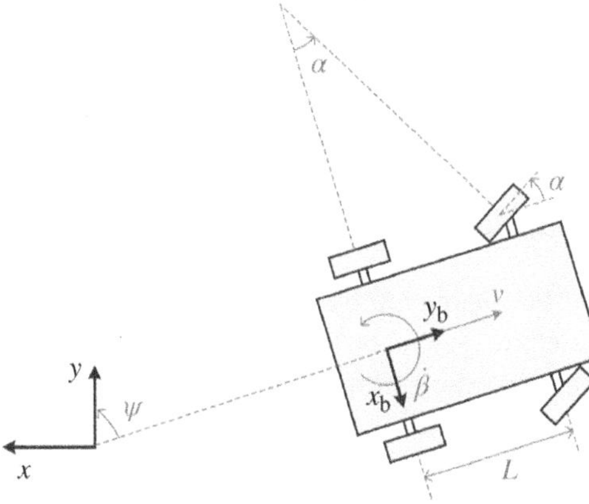

Figure 6.1 Reference frames and the steering model, where x, y denote the x and y axes of the global frame; x_b, y_b denote the x and y axes of the body frame. The vehicle's forward velocity, heading, and rotational velocity are denoted as v, ψ, and $\dot{\beta}$, respectively. Notations α and L represent the steering wheel angle and front-rear wheel base.

moves translationally along the body y axis in 2D space and rotates around the body z axis, we choose as state variables

$$\mathbf{x} = \begin{bmatrix} x & y & v & \psi & \dot{\beta} \end{bmatrix}^{\mathrm{T}} \tag{6.1}$$

where x, y are vehicle coordinates in meters; ψ is the heading in rads; and v and $\dot{\beta}$ denote the projection of the translational velocity onto the body y axis (in m/s) and the projection of the rotational velocity onto the body z axis (in rad/s). The continuous system equations are

$$\frac{\mathrm{d}}{\mathrm{d}t}\begin{bmatrix} x \\ y \\ v \\ \psi \\ \dot{\beta} \end{bmatrix} = \begin{bmatrix} -v\sin\psi \\ v\cos\psi \\ 0 \\ \dot{\beta} \\ 0 \end{bmatrix}. \tag{6.2}$$

By denoting the time interval as Δt, the discrete kinematic equation can be derived as

$$\mathbf{x}_{k+1} = \mathbf{A}(k)\mathbf{x}_k + \mathbf{B}\mathbf{h}(\mathbf{u}_k) + \mathbf{w}_k, \tag{6.3}$$

where $\mathbf{w}_k \sim \mathcal{N}(\mathbf{0}, \mathbf{Q})$ denotes independent and identically distributed (i.i.d.) Gaussian process noise with zero mean and covariance matrix $\mathbf{Q} > 0$. $\mathbf{A}(k)$ is written as

$$\mathbf{A}(k) = \begin{bmatrix} 1 & 0 & a_{13} & 0 & 0 \\ 0 & 1 & a_{23} & 0 & 0 \\ 0 & 0 & 1 & 0 & 0 \\ 0 & 0 & 0 & 1 & a_{45} \\ 0 & 0 & 0 & 0 & 1 \end{bmatrix} \tag{6.4}$$

$$\mathbf{B} = \begin{bmatrix} 0 & 0 & 1 & 0 & 0 \\ 0 & 0 & 0 & 0 & 1 \end{bmatrix}^{\mathrm{T}} \tag{6.5}$$

in which $a_{13} = -\sin\hat{\psi}_k \Delta t$, $a_{23} = \cos\hat{\psi}_k \Delta t$, $a_{45} = \Delta t$; $\hat{\psi}_k$ denotes the newest estimated heading; $\mathbf{u} = [u_v, u_{\dot{\beta}}]^\mathrm{T}$ is the control vector, where subvectors denote translational acceleration in $\mathrm{m/s^2}$ and angular acceleration in $\mathrm{rad/s^2}$, respectively. $\mathbf{h}(\mathbf{u}_k)$ is a stochastic function satisfying $\mathbb{E}\{\mathbf{h}(\mathbf{u}_k)|\mathbf{x}_k\} = 0$ for all $\mathbf{x}_k$.

6.3.2 Measurement Model

We consider a multi-sensor measurement model that includes a global navigation satellite system (GNSS), an attitude and heading reference system (AHRS), and encoders measuring the translational velocity and steering angle. In a loosely coupled model where GNSS measurements are converted from an earth-centered, earth-fixed frame to the global frame, the measurement equations of a dead-reckoning system can be written as [Kelly, 2013]

$$\mathbf{z} = \mathbf{G}\mathbf{x} + \mathbf{v} \tag{6.6}$$

where $\mathbf{z} = [\mathbf{z}_{\mathrm{GNSS}}^\mathrm{T} \in \mathbb{R}^2, \mathbf{z}_{\mathrm{AHRS}}^\mathrm{T} \in \mathbb{R}^1, \mathbf{z}_{\mathrm{ENC}}^\mathrm{T} \in \mathbb{R}^2]^\mathrm{T}$, $\mathbf{v} \sim \mathcal{N}(\mathbf{0}, \mathbf{R})$ is i.i.d. measurement Gaussian noise with zero mean and covariance matrix $\mathbf{R} > 0$. We note that the measurement equation for the steering angle is non-linear as $\alpha = \tan^{-1}\left(\frac{L\dot{\beta}}{v}\right)$ where the wheelbase between the front and rear wheels is denoted by L. By taking derivatives with regard to the state vector, the measurement matrix can be obtained as

$$\mathbf{G} = \begin{bmatrix} 1 & 0 & 0 & 0 & 0 \\ 0 & 1 & 0 & 0 & 0 \\ 0 & 0 & 0 & 1 & 0 \\ 0 & 0 & 1 & 0 & 0 \\ 0 & 0 & \frac{\partial\alpha}{\partial v} & 0 & \frac{\partial\alpha}{\partial\dot{\beta}} \end{bmatrix} \tag{6.7}$$

and $\frac{\partial\alpha}{\partial v} = \frac{-L\dot{\beta}}{v^2+(L\dot{\beta})^2}$, $\frac{\partial\alpha}{\partial\dot{\beta}} = \frac{Lv}{v^2+(L\dot{\beta})^2}$. All timing index subscripts in this measurement model have been omitted for convenient representation.

6.3.3 Attack Model

In this chapter, we assume that the sensors are subject to randomly occurring deception attacks with a given probability. The attack model is described as follows

$$\tilde{\mathbf{z}}_k = \mathbf{z}_k + \gamma_k \mathbf{a}_k = \mathbf{G}\mathbf{x}_k + \mathbf{v}_k + \gamma_k \mathbf{a}_k \tag{6.8}$$

where $\tilde{\mathbf{z}}_k$ denotes the measurement with possible attacks, $\mathbf{a}_k$ denotes the information sent by attacks, and γ_k is a stochastic variable.

Obviously, we can verify that $(\mathbf{A}(k), \mathbf{G})$ is observable for all k. Then according to the corresponding results of Fawzi et al. (2014a) and Mo and Garone (2016) for secure state estimation, we still need to assume that fewer than half of the pose states are compromised by attacks in this chapter. As a result, we can give a stable pose estimation for the AV in the presence of attacks.

The information $\mathbf{a}_k$ caused by deception attacks can be regarded as $\mathbf{a}_k = -\mathbf{z}_k + \xi_k$, where the nonzero ξ_k satisfying $||\xi_k|| \leq \delta$ is an arbitrary energy-bounded signal. The stochastic variable γ_k is a Bernoulli-distributed white sequence taking values of $\{0, 1\}$ with probabilities

$$\Pr\{\gamma_k = 0\} = \bar{\gamma}, \ \Pr\{\gamma_k = 1\} = 1 - \bar{\gamma}.$$

where $\bar{\gamma} \in (0, 1]$ is a known constant.

Remark 6.1 It should be mentioned that sensor attacks on AVs are considered a type of deception attack in our chapter. Taking account of the application of safety protection devices for AVs, there is a nonzero probability for each attack (e.g. spoofing attack) to be unsuccessful at a certain time. The model proposed in (6.8) can describe randomly occurring deception attacks: that is, the stochastic variable γ_k is utilized to govern the random nature of sensor attacks on AVs. Although we do not allow $\bar{\gamma} = 0$ in the proposed approach, we could approximate the systematic attack by letting $\bar{\gamma}$ be close to zero.

Remark 6.2 The false data sent by deception attackers could be identified using algorithms (e.g. χ^2 detectors) and some hardware and software tools. According to the definition of frequentist probability, we may deduce the value of $\bar{\gamma}$ in applications. Hence, the given Bernoulli distribution can properly reveal the random nature of deception attacks.

6.4 Estimator Design

In this section, we design a Kalman-type secure recursive estimator as follows

$$\hat{\mathbf{x}}_{k+1|k} = \mathbf{A}(k)\hat{\mathbf{x}}_k + \mathbf{B}\mathbf{h}(\mathbf{u}_k) \tag{6.9a}$$

$$\hat{\mathbf{x}}_{k+1} = \hat{\mathbf{x}}_{k+1|k} + \mathbf{K}_{k+1}(\tilde{\mathbf{z}}_{k+1} - \bar{\gamma}\mathbf{G}\hat{\mathbf{x}}_{k+1|k}) \tag{6.9b}$$

where $\hat{\mathbf{x}}_k$ is the state estimate at time k, $\hat{\mathbf{x}}_{k+1|k}$ is the one-step prediction at time k, and $\mathbf{K}_{k+1}$ is the estimator parameter matrix to be determined by the time $k + 1$.

Remark 6.3 It is not difficult to see that the Kalman-type secure recursive estimator is identical to the standard Kalman filter when $\bar{\gamma} = 1$, i.e. $\gamma_k = 0$ with probability 1 for all k. As a result, when the AV is not under attack with probability 1, our proposed estimator can recover the optimal Kalman estimate with probability 1.

Now we present the following lemmas, which will be used to derive our recursive estimator parameter matrix $\mathbf{K}_{k+1}$ in this chapter.

Lemma 6.1 *[Chen and Zheng, 2007a] For any dimension-compatible matrices* $\mathbf{D}$, $\mathbf{E}$, *and a scalar* $\varepsilon > 0$, *the following inequality holds:*

$$\mathbf{DE} + \mathbf{E}^{\mathrm{T}}\mathbf{D}^{\mathrm{T}} \leq \varepsilon\mathbf{DD}^{\mathrm{T}} + \varepsilon^{-1}\mathbf{EE}^{\mathrm{T}}$$

Lemma 6.2 *[Theodor and Shaked, 1996] Let* $\mathbf{f}_k(\cdot) : \mathbb{R}^{n\times n} \rightarrow \mathbb{R}^{n\times n}, 0 \leq k < N$, *be a sequence of matrix functions such that*

$$\mathbf{f}_k(\mathbf{U}) = \mathbf{f}_k^{\mathrm{T}}(\mathbf{U}),\ \forall\ \mathbf{U} = \mathbf{U}^{\mathrm{T}} > 0,$$
$$\mathbf{f}_k(\mathbf{V}) \geq \mathbf{f}_k(\mathbf{U}),\ \forall\ \mathbf{V} = \mathbf{V}^{\mathrm{T}} > \mathbf{U} = \mathbf{U}^{\mathrm{T}},$$

and let $\mathbf{g}_k(\cdot) : \mathbb{R}^{n\times n} \rightarrow \mathbb{R}^{n\times n}, 0 \leq k < N$, *be a sequence of matrix functions such that*

$$\mathbf{g}_k(\mathbf{U}) = \mathbf{g}_k^{\mathrm{T}}(\mathbf{U}) \geq \mathbf{f}_k(\mathbf{U}),\ \forall\ \mathbf{U} = \mathbf{U}^{\mathrm{T}} > 0.$$

Then the solutions $\{\mathbf{U}_k\}_{0\leq k<N}$ *and* $\{\mathbf{V}_k\}_{0\leq k<N}$ *to the difference equations*

$$\mathbf{U}_{k+1} = \mathbf{f}_k(\mathbf{U}_k),\ \mathbf{V}_{k+1} = \mathbf{g}_k(\mathbf{V}_k),\ \mathbf{U}_0 = \mathbf{V}_0 > 0,$$

satisfy $\mathbf{U}_k \leq \mathbf{V}_k$ *for* $0 \leq k < N$.

The objective of this chapter is to design a finite-horizon estimator of (6.9a) and (6.9b) such that, for randomly occurring deception attacks, an upper bound for the estimation error covariance is guaranteed: that is, there exists a sequence of positive-definite matrices $\mathbf{\Pi}_{k+1}$ $(0 \leq k \leq N)$ satisfying

$$\mathbb{E}\{(\mathbf{x}_{k+1} - \hat{\mathbf{x}}_{k+1})(\mathbf{x}_{k+1} - \hat{\mathbf{x}}_{k+1})^{\mathrm{T}}\} \leq \mathbf{\Pi}_{k+1},\ \forall k.$$

Moreover, the designed estimator parameter matrix $\mathbf{K}_{k+1}$ is expected to minimize the upper bound $\mathbf{\Pi}_{k+1}$ through a recursive scheme.

Define $\mathbf{\Sigma}_{k+1} = \mathbb{E}\{\mathbf{x}_{k+1}\mathbf{x}_{k+1}^{\mathrm{T}}\}$; then we know that

$$\begin{aligned}\mathbf{\Sigma}_{k+1} &= \mathbb{E}\{[\mathbf{A}(k)\mathbf{x}_k + \mathbf{Bh}(\mathbf{u}_k) + \mathbf{w}_k][\mathbf{A}(k)\mathbf{x}_k + \mathbf{Bh}(\mathbf{u}_k) + \mathbf{w}_k]^{\mathrm{T}}\} \\ &= \mathbf{A}(k)\mathbf{\Sigma}_k\mathbf{A}^{\mathrm{T}}(k) + \mathbf{Bh}(\mathbf{u}_k)\mathbf{h}^{\mathrm{T}}(\mathbf{u}_k)\mathbf{B}^{\mathrm{T}} + \mathbf{Q}\end{aligned} \tag{6.10}$$

Denote the one-step prediction error and the estimation error as $\mathbf{e}_{k+1|k} = \mathbf{x}_{k+1} - \hat{\mathbf{x}}_{k+1|k}$ and $\mathbf{e}_k = \mathbf{x}_k - \hat{\mathbf{x}}_k$, respectively. So the one-step prediction error covariance matrix $\mathbf{P}_{k+1|k}$ and the estimation error covariance matrix $\mathbf{P}_{k+1}$ can be obtained as follows:

$$\mathbf{P}_{k+1|k} = \mathbb{E}\{[\mathbf{x}_{k+1} - \hat{\mathbf{x}}_{k+1|k}][\mathbf{x}_{k+1} - \hat{\mathbf{x}}_{k+1|k}]^{\mathrm{T}}\}, \tag{6.11a}$$

$$\mathbf{P}_{k+1} = \mathbb{E}\{[\mathbf{x}_{k+1} - \hat{\mathbf{x}}_{k+1}][\mathbf{x}_{k+1} - \hat{\mathbf{x}}_{k+1}]^{\mathrm{T}}\}. \tag{6.11b}$$

Now let us show the unbiasedness of recursive estimator (6.9a) and (6.9b) with $\xi_k = 0$.

$$\begin{aligned}\mathbb{E}\{\mathbf{e}_{k+1}\} &= \mathbb{E}\{(\mathbf{x}_{k+1} - \hat{\mathbf{x}}_{k+1})\}\\ &= [\mathbf{I} - (1-\bar{\gamma})\mathbf{K}_{k+1}\mathbf{G}]\mathbf{A}(k)\mathbb{E}\{\mathbf{e}_k\} \qquad (6.12)\end{aligned}$$

From the initial value $\hat{\mathbf{x}}_0 = \mathbb{E}\{\mathbf{x}_0\}$, the recursive estimator (6.9a) and (6.9b) with $\xi_k = 0$ is an unbiased estimator.

We are now ready to conduct the one-step prediction error matrix in terms of the solvability of recursive Riccati difference equations and obtain the parameter gain matrix of Kalman-type recursive estimator (6.9a) and (6.9b), which is developed in the following theorem.

Theorem 6.1 *Consider the discrete kinematic equation* (6.3) *suffering from attacks as* (6.8). *For any given positive constants* $\varepsilon_k, k = 0, 1, 2, \cdots$, *and the initial condition* $\mathbf{x}_0$, $\hat{\mathbf{x}}_0 = \mathbb{E}\{\mathbf{x}_0\}$, $\mathbf{\Pi}_0 = \mathbf{P}_0$, $\mathbf{\Sigma}_0 = \mathbb{E}\{\mathbf{x}_0\mathbf{x}_0^{\top}\}$, *we can derive that the parameter gain matrix of Kalman-type recursive estimator* (6.9a) *and* (6.9b) *is given as follows*

$$\begin{aligned}\mathbf{K}_{k+1} &= \alpha_4\mathbf{\Pi}_{k+1|k}\mathbf{G}^{\top}[\alpha_1\mathbf{G}\mathbf{\Pi}_{k+1|k}\mathbf{G}^{\top} + \alpha_2\mathbf{I}\\ &\quad + \bar{\gamma}\mathbf{R} + \alpha_3\mathbf{G}\mathbf{\Sigma}_{k+1}\mathbf{G}^{\top}]^{-1}, \qquad (6.13)\end{aligned}$$

where $\alpha_1 = [1 + (1-\bar{\gamma})\varepsilon_k]\bar{\gamma}^2$, $\alpha_2 = (1-\bar{\gamma}^2)\varepsilon_k^{-1}\delta^2 + (1-\bar{\gamma})\delta^2$, $\alpha_3 = \bar{\gamma}(1-\bar{\gamma})(1+\varepsilon_k)$, *and* $\alpha_4 = \bar{\gamma} + (\bar{\gamma} - \bar{\gamma}^2)\varepsilon_k$. *The upper bound for the estimation error covariance is* $\mathbf{\Pi}_{k+1}$, *which can be recursively calculated by the equation* (6.19b).

Proof: From (6.11a), we know that

$$\begin{aligned}\mathbf{P}_{k+1|k} &= \mathbb{E}\{[\mathbf{A}(k)\mathbf{e}_k + \mathbf{w}_k][\mathbf{A}(k)\mathbf{e}_k + \mathbf{w}_k]^{\top}\}\\ &= \mathbf{A}(k)\mathbf{P}_k\mathbf{A}^{\top}(k) + \mathbf{Q}. \qquad (6.14)\end{aligned}$$

Since $\mathbf{e}_k = \mathbf{x}_k - \hat{\mathbf{x}}_k$, it can be obtained that

$$\begin{aligned}\mathbf{e}_{k+1} &= \mathbf{x}_{k+1} - [\hat{\mathbf{x}}_{k+1|k} + \mathbf{K}_{k+1}(\tilde{\mathbf{z}}_{k+1} - \bar{\gamma}\mathbf{G}\hat{\mathbf{x}}_{k+1|k})]\\ &= \mathbf{x}_{k+1} - \hat{\mathbf{x}}_{k+1|k} - \bar{\gamma}\mathbf{K}_{k+1}\mathbf{G}(\mathbf{x}_{k+1} - \hat{\mathbf{x}}_{k+1|k})\\ &\quad + \mathbf{K}_{k+1}(\bar{\gamma}\mathbf{G}\mathbf{x}_{k+1} - \tilde{\mathbf{z}}_{k+1})\\ &= (\mathbf{I} - \bar{\gamma}\mathbf{K}_{k+1}\mathbf{G})\mathbf{e}_{k+1|k} - \mathbf{K}_{k+1}(\gamma_{k+1}\mathbf{a}_{k+1}\\ &\quad + \mathbf{v}_{k+1} + (1-\bar{\gamma})\mathbf{G}\mathbf{x}_{k+1}). \qquad (6.15)\end{aligned}$$

From (6.11b), we know that

$$\begin{aligned}\mathbf{P}_{k+1} &= \mathbb{E}\{[(\mathbf{I} - \bar{\gamma}\mathbf{K}_{k+1}\mathbf{G})\mathbf{e}_{k+1|k} - \mathbf{K}_{k+1}(\gamma_{k+1}\mathbf{a}_{k+1}\\ &\quad + \mathbf{v}_{k+1} + (1-\bar{\gamma})\mathbf{G}\mathbf{x}_{k+1})][(\mathbf{I} - \bar{\gamma}\mathbf{K}_{k+1}\mathbf{G})\mathbf{e}_{k+1|k}\\ &\quad - \mathbf{K}_{k+1}(\gamma_{k+1}\mathbf{a}_{k+1} + \mathbf{v}_{k+1} + (1-\bar{\gamma})\mathbf{G}\mathbf{x}_{k+1})]^{\top}\}. \qquad (6.16)\end{aligned}$$

Then we can obtain that

$$\begin{aligned}\mathbf{P}_{k+1} &= (\mathbf{I}-\bar{\gamma}\mathbf{K}_{k+1}\mathbf{G})\mathbf{P}_{k+1|k}(\mathbf{I}-\bar{\gamma}\mathbf{K}_{k+1}\mathbf{G})^{\top}\\ &\quad-(1-\bar{\gamma})(\mathbf{I}-\bar{\gamma}\mathbf{K}_{k+1}\mathbf{G})\mathbb{E}\{\mathbf{e}_{k+1|k}\xi_{k+1}^{\top}\}\mathbf{K}_{k+1}^{\top}\\ &\quad-(1-\bar{\gamma})\mathbf{K}_{k+1}\mathbb{E}\{\xi_{k+1}\mathbf{e}_{k+1|k}^{\top}\}(\mathbf{I}-\bar{\gamma}\mathbf{K}_{k+1}\mathbf{G})^{\top}\\ &\quad-\bar{\gamma}(1-\bar{\gamma})\mathbf{K}_{k+1}\mathbf{G}\mathbb{E}\{\mathbf{x}_{k+1}\xi_{k+1}^{\top}\}\mathbf{K}_{k+1}^{\top}\\ &\quad-\bar{\gamma}(1-\bar{\gamma})\mathbf{K}_{k+1}\mathbb{E}\{\xi_{k+1}\mathbf{x}_{k+1}^{\top}\}\mathbf{G}^{T}\mathbf{K}_{k+1}^{\top}\\ &\quad+\bar{\gamma}(1-\bar{\gamma})\mathbf{K}_{k+1}\mathbf{G}\mathbf{\Sigma}_{k+1}\mathbf{G}^{\top}\mathbf{K}_{k+1}^{\top}\\ &\quad+(1-\bar{\gamma})\mathbf{K}_{k+1}\xi_{k+1}\xi_{k+1}^{\top}\mathbf{K}_{k+1}^{\top}\\ &\quad+\bar{\gamma}\mathbf{K}_{k+1}\mathbf{R}\mathbf{K}_{k+1}^{\top}. \end{aligned} \tag{6.17}$$

By **Lemma** 6.1 and applying the property of matrix trace, we have that

$$\begin{aligned}\mathbf{P}_{k+1} &\le [1+(1-\bar{\gamma})\varepsilon_k](\mathbf{I}-\bar{\gamma}\mathbf{K}_{k+1}\mathbf{G})\mathbf{P}_{k+1|k}(\mathbf{I}-\bar{\gamma}\mathbf{K}_{k+1}\mathbf{G})^{\top}\\ &\quad+[\bar{\gamma}(1-\bar{\gamma})(1+\varepsilon_k)]\mathbf{K}_{k+1}\mathbf{G}\mathbf{\Sigma}_{k+1}\mathbf{G}^{\top}\mathbf{K}_{k+1}^{\top}\\ &\quad+[(1-\bar{\gamma}^2)\varepsilon_k^{-1}\delta^2+(1-\bar{\gamma})\delta^2]\mathbf{K}_{k+1}\mathbf{K}_{k+1}^{\top}\\ &\quad+\bar{\gamma}\mathbf{K}_{k+1}\mathbf{R}\mathbf{K}_{k+1}^{\top}. \end{aligned} \tag{6.18}$$

Define

$$\mathbf{\Pi}_{k+1|k} = \mathbf{A}(k)\mathbf{\Pi}_k\mathbf{A}^{\top}(k)+\mathbf{Q}, \tag{6.19a}$$

$$\begin{aligned}\mathbf{\Pi}_{k+1} &= [1+(1-\bar{\gamma})\varepsilon_k](\mathbf{I}-\bar{\gamma}\mathbf{K}_{k+1}\mathbf{G})\mathbf{\Pi}_{k+1|k}(\mathbf{I}-\bar{\gamma}\mathbf{K}_{k+1}\mathbf{G})^{\top}\\ &\quad+[\bar{\gamma}(1-\bar{\gamma})(1+\varepsilon_k)]\mathbf{K}_{k+1}\mathbf{G}\mathbf{\Sigma}_{k+1}\mathbf{G}^{\top}\mathbf{K}_{k+1}^{\top}\\ &\quad+[(1-\bar{\gamma}^2)\varepsilon_k^{-1}\delta^2+(1-\bar{\gamma})\delta^2]\mathbf{K}_{k+1}\mathbf{K}_{k+1}^{\top}\\ &\quad+\bar{\gamma}\mathbf{K}_{k+1}\mathbf{R}\mathbf{K}_{k+1}^{\top}. \end{aligned} \tag{6.19b}$$

Applying **Lemma** 6.2 to (6.14), (6.18), (6.19a) and (6.19b), we can obtain that

$$\mathbf{P}_{k+1|k} \le \mathbf{\Pi}_{k+1|k},\ \mathbf{P}_{k+1} \le \mathbf{\Pi}_{k+1}. \tag{6.20}$$

Taking the partial derivation of the trace of the matrix $\mathbf{\Pi}_{k+1}$ with respect to $\mathbf{K}_{k+1}$, and letting the derivative be zero, we can obtain that

$$\begin{aligned}&-2[\bar{\gamma}+(\bar{\gamma}-\bar{\gamma}^2)\varepsilon_k](\mathbf{I}-\bar{\gamma}\mathbf{K}_{k+1}\mathbf{G})\mathbf{\Pi}_{k+1|k}\mathbf{G}^{\top}\\ &+2[(1-\bar{\gamma}^2)\varepsilon_k^{-1}\delta^2+(1-\bar{\gamma})\delta^2]\mathbf{K}_{k+1}\\ &+2\bar{\gamma}(1-\bar{\gamma})(1+\varepsilon_k)\mathbf{K}_{k+1}\mathbf{G}\mathbf{\Sigma}_{k+1}\mathbf{G}^{\top}+2\bar{\gamma}\mathbf{K}_{k+1}\mathbf{R}=0. \end{aligned} \tag{6.21}$$

It follows that

$$\begin{aligned}&\mathbf{K}_{k+1}[\alpha_1\mathbf{G}\mathbf{\Pi}_{k+1|k}\mathbf{G}^\top+\alpha_2\mathbf{I}+\overline{\gamma}\mathbf{R}+\alpha_3\mathbf{G}\mathbf{\Sigma}_{k+1}\mathbf{G}^\top]\\&\quad=\alpha_4\mathbf{\Pi}_{k+1|k}\mathbf{G}^\top.\end{aligned}$$

Since $\alpha_1\mathbf{G}\mathbf{\Pi}_{k+1|k}\mathbf{G}^\top+\alpha_2\mathbf{I}+\overline{\gamma}\mathbf{R}+\alpha_3\mathbf{G}\mathbf{\Sigma}_{k+1}\mathbf{G}^\top$ is a positive definite matrix, we know (6.13) holds and the proof is complete. □

As we can see in (6.13) and (6.14), we need $\mathcal{O}(n^3)$ operations to compute the Kalman gain and the covariance matrix $P_{k+1|k}$, where $n = 5$ in this chapter. This indicates that the Kalman-type secure recursive estimator can be treated in a short time with a high-performance computer.

Remark 6.4 It should be mentioned that the phenomenon of randomly occurring deception attacks (sensor attacks) is the first concern for AVs in our chapter. In **Theorem 1**, a secure pose estimator for AVs has been proposed to minimize the upper bound of the estimation error covariance, which can be solved very efficiently in real time and is suitable for recursive computation in online applications.

Remark 6.5 From (6.19b), it can be seen that the larger δ leads to a bigger upper bound of the estimation error covariance, which means the estimation performance deteriorates with increased δ.

Remark 6.6 According to the matrix inequality technique of **Lemma** 6.1, we can arbitrary choose the positive constant ε_k in **Theorem 1** from a theoretical point of view. However, a too-large or too-small value of ε_k may influence the estimation performance. In practice or experimental validation, we select the appropriate positive constant ε_k based on experience to achieve better estimation performance.

As a matter of fact, for AVs in the presence of deception attacks (sensor attacks), how to obtain a secure pose estimation of the discrete kinematic equation (6.3) has remained an open problem until now. The goal of this chapter is to propose an algorithm that enables to estimate of the pose states of the AV in such a way that:

1. If no sensors (GNSS and AHRS) are comprised, i.e. $\overline{\gamma} = 1$ and $\gamma_k = 0$ with probability 1 for all k, the estimate coincides with the standard Kalman filter.
2. If fewer than half of the pose states are compromised by randomly occurring deception attacks, the algorithm still gives a stable estimate of the pose states: i.e. an upper bound for the estimation error covariance is guaranteed.

According to **Theorem** 6.1, the calculation framework can be summarized as the following algorithm.

Algorithm 4 Kalman-type secure recursive estimator

Step 1. Initialization:

1. Give the values of initial pose state $\mathbf{x}_0$, initial estimate state $\hat{\mathbf{x}}_0$, initial estimation error covariance matrix $\mathbf{P}_0$, and initial state covariance matrix $\mathbf{\Sigma}_0$;
2. Give the value of $\bar{\gamma}$, and determine the value of δ (the norm bound of arbitrary signal ξ_k);
3. Give the control input signal $\mathbf{u}_k$, i.e. translational and rotational velocities for the AV;
4. Let $\mathbf{\Pi}_0 = \mathbf{P}_0$, and choose the proper ε_k for all k to calculate α_1, α_2, α_3, and α_4.

Step 2. For $k=0$, the state covariance matrix $\mathbf{\Sigma}_{k+1}$ and $\mathbf{\Pi}_{k+1|k}$ are updated as follows:

$$\mathbf{\Sigma}_{k+1} = \mathbf{A}(k)\mathbf{\Sigma}_k\mathbf{A}^{\mathrm{T}}(k) + \mathbf{B}\mathbf{h}(\mathbf{u}_k)\mathbf{h}^{\mathrm{T}}(\mathbf{u}_k)\mathbf{B}^{\mathrm{T}} + \mathbf{Q}$$
$$\mathbf{\Pi}_{k+1|k} = \mathbf{A}(k)\mathbf{\Pi}_k\mathbf{A}^{\mathrm{T}}(k) + \mathbf{Q}$$

Step 3. For $k=0$, the Kalman-type secure recursive estimator gain $\mathbf{K}_{k+1}$ and $\mathbf{\Pi}_{k+1}$ are calculated as follows:

$$\begin{aligned}\mathbf{K}_{k+1} &= \alpha_4\mathbf{\Pi}_{k+1|k}\mathbf{G}^{\mathrm{T}}[\alpha_1\mathbf{G}\mathbf{\Pi}_{k+1|k}\mathbf{G}^{\mathrm{T}} + \alpha_2\mathbf{I} \\ &\quad + \bar{\gamma}\mathbf{R} + \alpha_3\mathbf{G}\mathbf{\Sigma}_{k+1}\mathbf{G}^{\mathrm{T}}]^{-1} \\ \mathbf{\Pi}_{k+1} &= [1+(1-\bar{\gamma})\varepsilon_k](\mathbf{I}-\bar{\gamma}\mathbf{K}_{k+1}\mathbf{G})\mathbf{\Pi}_{k+1|k}(\mathbf{I}-\bar{\gamma}\mathbf{K}_{k+1}\mathbf{G})^{\mathrm{T}} \\ &\quad + [\bar{\gamma}(1-\bar{\gamma})(1+\varepsilon_k)]\mathbf{K}_{k+1}\mathbf{G}\mathbf{\Sigma}_{k+1}\mathbf{G}^{\mathrm{T}}\mathbf{K}_{k+1}^{\mathrm{T}} \\ &\quad + [(1-\bar{\gamma}^2)\varepsilon_k^{-1}\delta^2 + (1-\bar{\gamma})\delta^2]\mathbf{K}_{k+1}\mathbf{K}_{k+1}^{\mathrm{T}} \\ &\quad + \bar{\gamma}\mathbf{K}_{k+1}\mathbf{R}\mathbf{K}_{k+1}^{\mathrm{T}}.\end{aligned}$$

Step 4. Set $k=k+1$, and go to Step 2.

6.5 Discussion of Parameter Selection

For practical implementations, the influence of parameter selection on estimation performance needs to be discussed such that researchers and practitioners are guided in parameter tuning.

6.5.1 The Probability Subject to Deception Attacks

As discussed earlier, if $\bar{\gamma} = 1$, i.e. no sensors (GNSS and AHRS) are comprised with probability 1, the Kalman-type secure recursive estimator is identical to the standard Kalman filter, which results in an optimal Kalman estimate. However, if $\bar{\gamma} \in (0, 1)$, i.e. the sensors (GNSS and AHRS) are subject to deception attacks with probability $1 - \bar{\gamma}$, then the Kalman-type secure recursive estimator may not be the optimal estimator but can give a stable estimate where the upper bound for the estimation error covariance is guaranteed.

Table 6.1 Parameters in the secure filter.

$\overline{\gamma}$	$\Pr\{\gamma_k = 0\} = \overline{\gamma} \in (0, 1]$, i.e. deception attacks with probability $1 - \overline{\gamma}$, smaller $\overline{\gamma}$ will lead to a bigger upper bound of the estimation error covariance.
δ	δ is the bound of an arbitrary energy signal sent by the adversary. A larger δ will lead to a bigger upper bound of the estimation error covariance.
ε_k	ε_k is an adjustable positive constant. Select a properly positive constant ε_k based on experience to achieve better estimation performance. The details can be seen in Remark 5.

From (6.19b), it can be seen that the value of $\overline{\gamma}$ will influence the value of the upper bound for the estimation error covariance. We have an instinctive feeling that a greater likelihood of suffering from attacks (i.e. larger $1 - \overline{\gamma}$ or smaller $\overline{\gamma}$) will lead to a bigger upper bound for the estimation error covariance. Since this result cannot be directly deduced from (6.19b), in our chapter we will let $\overline{\gamma} = \{0.1, 0.2, 0.3, 0.4, 0.5, 0.6, 0.7, 0.8, 0.9, 1\}$ to simulate and compare with the estimation performance in the experimental validation.

6.5.2 The Bound of Signal ξ_k

For deception attacks, the adversary sends false information $\mathbf{a}_k = -\mathbf{z}_k + \xi_k$ to controllers or actuators at time k, where $\|\xi_k\| \leq \delta$ are the arbitrary bounded energy signals sent by the adversary. Obviously, if $\delta = 0$, the signal ξ_k will also be zero for all k and $\mathbf{a}_k = -\mathbf{z}_k$. From (6.8), we know that the case $\delta = 0$ can be regarded as sensors failure $\tilde{\mathbf{z}}_k = 0$ caused by the attack with probability $1 - \overline{\gamma}$.

A larger δ leads to a bigger upper bound of the estimation error covariance, which means the estimation performance deteriorates with increased δ. Therefore, in the experimental validation, we cannot choose a large value of δ and should select relatively small values of δ to achieve better estimation performance.

Therefore, the parameters and their influences on estimation performance in the proposed secure filter are summarized in Table 6.1.

6.6 Experimental Validation

The proposed iterative estimator is convenient for real-time applications. To verify the effectiveness of the proposed state estimator, we have run extensive simulations under Matlab R2016a based on a previously created system model, measurement model, and attack model. The testing route and control signals are shown in Figure 6.2, where the control signal is designed to cover common

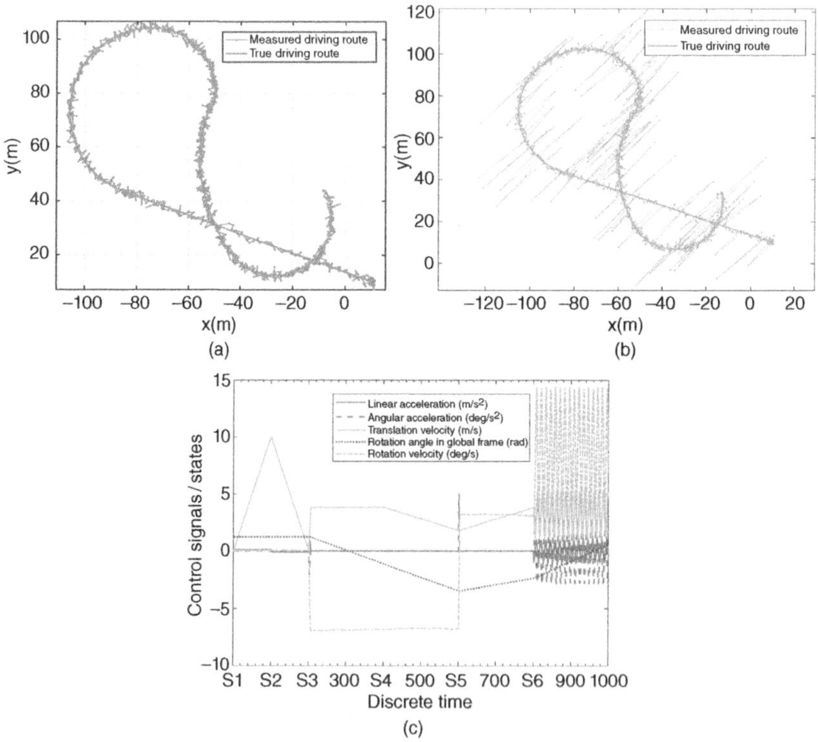

Figure 6.2 (a) The testing route with process noise and measurement noise only. (b) The testing route with noises and example attacks $\mathbf{a}_k = [20 \sin k, 20 \sin k, \mathbf{0}]^\mathsf{T}$ with probability 0.1, i.e. $\bar{\gamma} = 0.9$. (c) Control signals added to the system and some system states. The proposed testing route consists of six segments, which demonstrate typical scenarios for AVs. S1: driving straight, accelerating; S2: driving straight, braking; S3: turning right at constant velocity; S4: turning right, braking slightly; S5: turning left, accelerating slightly; S6: sinusoidal control signals.

scenarios including driving straight, making turns, accelerating, and braking. For all experiments, the initial estimation error covariance matrix $\mathbf{P}_0$ and initial state covariance matrix $\mathbf{\Sigma}_0$ are set to identity matrices. The initial state $\mathbf{x}_0 = [10, 10, 0.001, 1.222, 0]^\mathsf{T}$, while its estimation $\hat{\mathbf{x}}_0$ is set to the zero vector. The covariance matrices for system noise and measurement noise are set to

$$\mathbf{Q} = \mathrm{diag}[0.001^2, 0.001^2, 0.001^2, 0.0005^2, 0.0001^2],$$
$$\mathbf{R} = \mathrm{diag}[1.0^2, 1.0^2, 0.0175^2, 0.5^2, 0.0349^2].$$

The parameters are set based on experience and the sensors' performance. In this work, $\mathbf{R}$ is set as described since it is assumed that elements in the measurement vector are mutually independent, and the standard deviations of GNSS translational errors, AHRS rotational error, wheel encoder speed error, and steering encoder angular error are 1 m (in both x and y direction), 1 degree, 0.5 m/s, and 2 degrees, respectively.

We use the Kalman filter (KF) as a performance benchmark since the proposed estimator originates from and degenerates to the KF with specific parameter settings. It would be straightforward and fair to show performance enhancements under attacks compared with KF.

6.6.1 Pose Estimation under Attack on a Single State

When there is no attack, the proposed estimator degenerates to a conventional Kalman estimator according to Eq. (6.9). In this section, we investigate the estimator's performance under attack on a single state by making a comparison with a conventional KF. In the following group of simulations, the non-attack probability $\overline{\gamma}$ varies among 0.3, 0.6, and 0.9, and other parameters are set as $\epsilon = 0.001$, $\delta = 300$. Note that $\mathbf{a}$ is unknown in a practical implementation but is required for generating the attack signal. The attack signal is designed as $\mathbf{a}_k^1 = [\mathbf{0}, 200\cos k]^\top$.

The comparison state estimation results are shown in Figure 6.3. Compared to frequent-attack scenarios, the KF performs better if fewer attacks occur. For states x and ψ, although there are offsets in state estimation, the proposed estimator shows robustness compared to the KF, which is obviously greatly influenced by attacks. For state $\dot{\beta}$, our estimation errors are much smaller than the competitor regardless of the probability of attack.

The L2 norm of the estimator covariance matrix is shown at the top of Figure 6.7. The estimation error does not converge to zero with sustained attacks; thus, the figure shows that the KF fails to provide a reliable metric for estimation error when attacks exist. On the other hand, the proposed estimator provides us with a theoretically guaranteed upper bound for the estimator error covariance at each iteration.

Figure 6.4 shows details by examining the diagonal elements of $\mathbf{\Pi}_k$ for the estimated error variances of states x, ψ, and $\dot{\beta}$. The upper bound plays a major role in evaluating estimator performance, and it can be used as a metric to decide when the estimator output should be discarded. For example, Figure 6.4(b) indicates that estimation on ψ is not reliable, which accords with the middle row in Figure 6.3.

According to experiments with regard to different $\overline{\gamma}$, we know that for this particular system and parameters, as mentioned earlier, the smaller the probability of suffering from sensor attacks, the lower the upper bound that can be obtained. This phenomenon is interpretable as meaning that when fewer chance measurements are tampered with by attacks, the estimator may provide more reliable results – which implies a smaller error covariance norm. Unfortunately, it is hard to generalize this property to all systems and parameter settings mathematically.

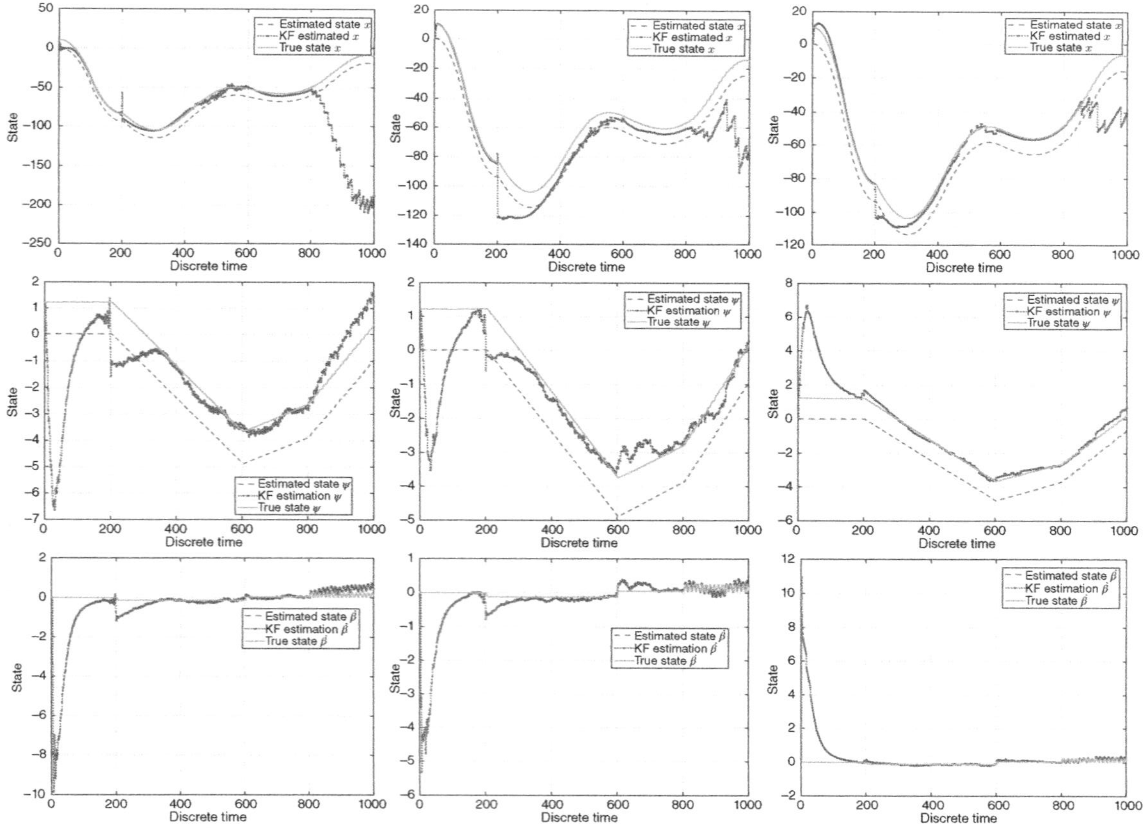

Figure 6.3 Estimator performance under attack $\mathbf{a}_k^1 = [\mathbf{0}, 200\cos k]^\top$ on single state. Upper, middle, and bottom rows show states x, ψ, and $\dot{\beta}$, respectively. Left, middle, and right columns illustrate results where $\overline{\gamma} = 0.3$, 0.6, and 0.9.

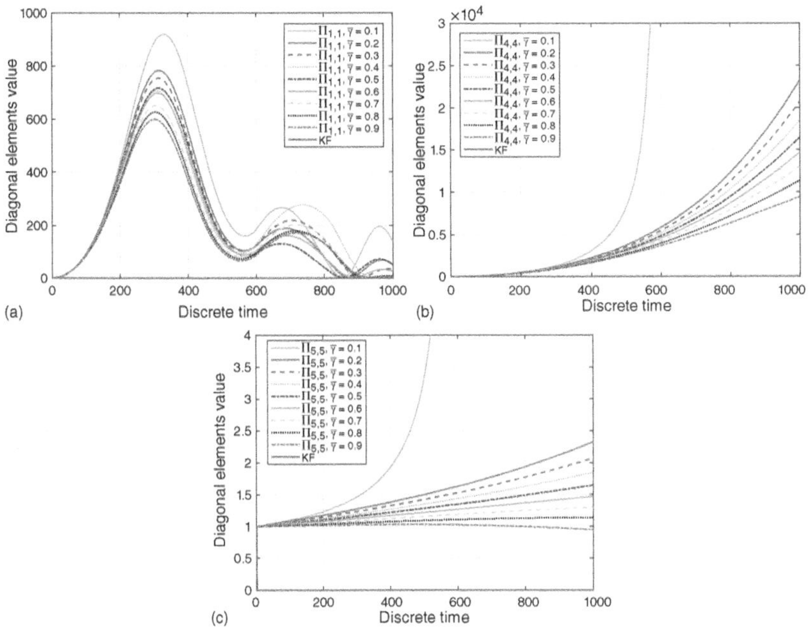

Figure 6.4 Diagonal elements (a) $\mathbf{\Pi}_{1,1}$; (b) $\mathbf{\Pi}_{4,4}$; and (c) $\mathbf{\Pi}_{5,5}$ with regard to different $\overline{\gamma}$.

6.6.2 Pose Estimation under Attacks on Multiple States

We next explore performance given sensor attacks on multiple measurement states. As discussed in Section 6.4, the estimator may not give stable output if over half of the measurement states are attacked. Herein, the attack signals on multiple states are selected as

$$\mathbf{a}_k^2 = \begin{bmatrix} 20\sin k & 20\sin k & 0 & 0 & 0 \end{bmatrix}^{\mathrm{T}} \tag{6.22}$$

$$\mathbf{a}_k^3 = \begin{bmatrix} 0 & 0 & 20\cos k & 0 & 20\cos k \end{bmatrix}^{\mathrm{T}} \tag{6.23}$$

Other parameters remain unchanged compared to the single-state attack case, expect for the bound of attack signals $\delta = 30$.

The state estimation results are shown in Figure 6.5 and Figure 6.6. The estimation becomes more accurate with lower attack probability. Specifically, while $\overline{\gamma} = 0.9$, there seems to be a converging tendency for estimation error; severe oscillations have been noticed under attacks $\mathbf{a}_k^3$ and $\overline{\gamma} = 0.3$.

The influence of attack probability on estimation accuracy can also be observed in Figure 6.7. As iterations of error covariance matrices are independent of attack signals, the graphs are identical for $\mathbf{a}_k^2$ and $\mathbf{a}_k^3$. Similar to a single-state attack case, the estimation becomes extremely insecure with decreasing $\overline{\gamma}$. When $\overline{\gamma} = 0.3$, the

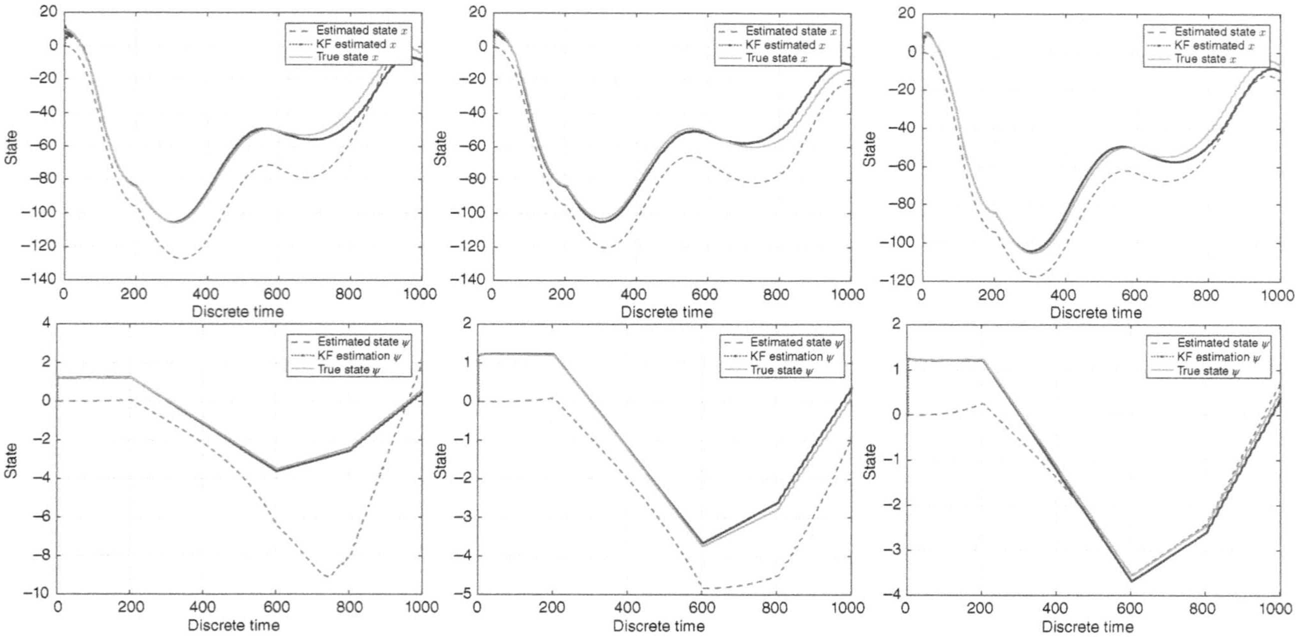

Figure 6.5 Estimator performance under attacks $\mathbf{a}_k^2 = [20 \sin k, 20 \sin k, \mathbf{0}]^\top$ on multiple states. The top and bottom rows show states x and ψ, respectively. The left, middle, and right columns illustrate results where $\overline{\gamma} = 0.3$, 0.6, and 0.9.

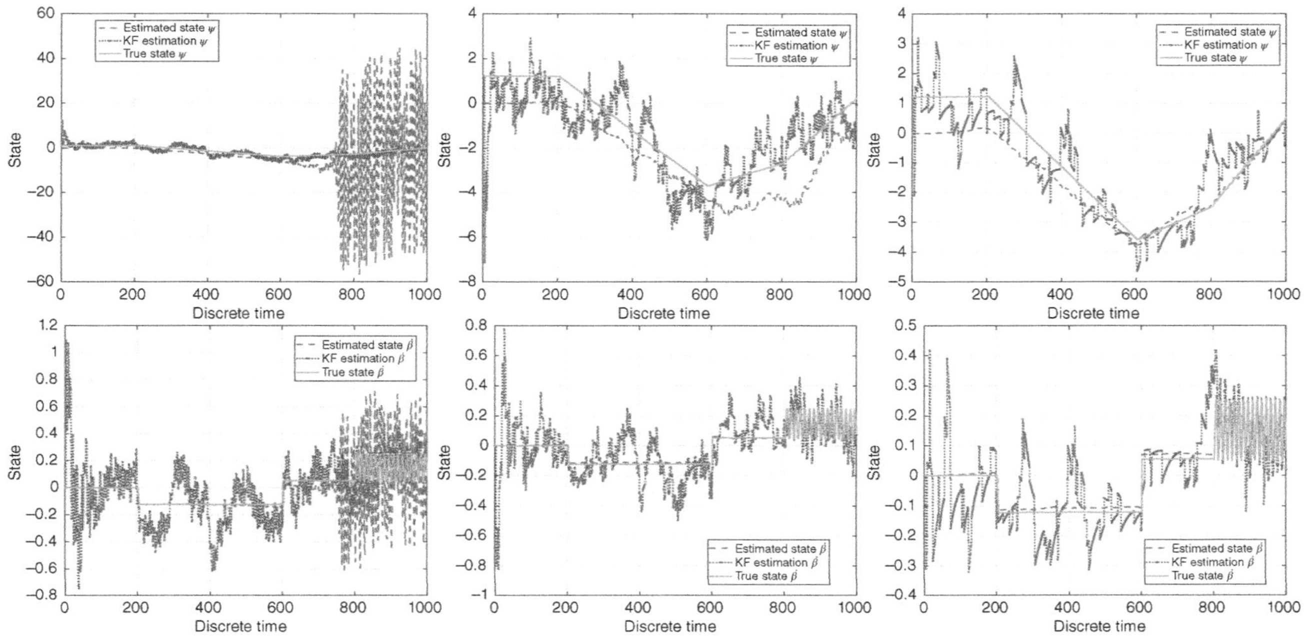

Figure 6.6 Estimator performance under attacks $\mathbf{a}_k^3 = [0, 0, 20\cos k, 0, 20\cos k]^{\top}$ on multiple states. The upper and bottom rows show states ψ and $\dot{\beta}$, respectively. The left, middle, and right columns illustrate results where $\overline{\gamma} = 0.3, 0.6,$ and 0.9.

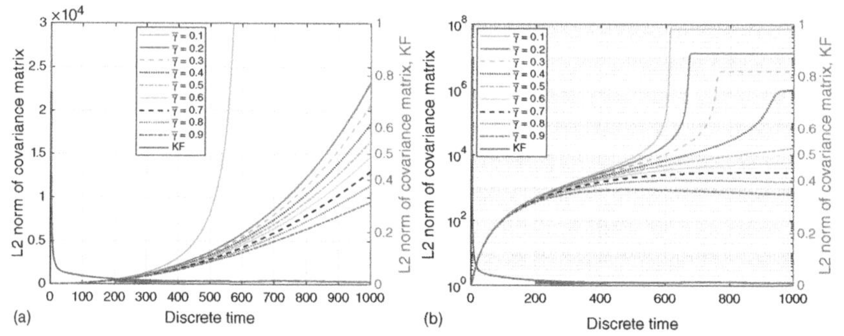

Figure 6.7 The L2 norm of the estimator error covariance matrix with regard to different $\bar{\gamma}$. Since iterations of the error covariance matrix rely only on the estimator parameters and are independent of attack signals, (a) illustrates the experiment with $\mathbf{a}_k^1$, and (b) demonstrates the situation when $\mathbf{a}_k^2$, $\mathbf{a}_k^3$ are added.

sudden step at $k \approx 750$ can be considered an indication that state estimation is highly unreliable.

From all the experimental results, we may conclude that under possible sensor attacks, the proposed estimator in this chapter provides a theoretically guaranteed upper bound for the error covariance matrix at the cost of forfeiting linear optimality compared to a conventional KF.

6.7 Conclusion

In this chapter, a recursive secure dynamic estimator has been designed to tackle the state estimation problem for autonomous vehicles in the circumstance of possible sensor attacks. We experimentally evaluated the proposed approach over multiple attack signals and showed that it outperforms conventional Kalman filtered results in the sense of offering an upper bound for the error covariance in addition to estimated states. In the future, it will be promising to implement the proposed approach on vehicle navigation systems to enable robust pose estimation. In addition, adaptive methods such as process noise scaling and reinforcement learning for parameter estimation may be developed to further enhance estimation performance.

7

UKF-Based Vehicle Pose Estimation under Randomly Occurring Deception Attacks

7.1 Introduction

With the continuous development of artificial intelligence (AI), the Internet of Things (IoT), and high-performance computing devices in intelligent transportation systems [Philip et al., 2018], autonomous vehicles (AVs) have become a focus of research in the last decade. Implemented with AV technologies, transportation safety and efficiency have been greatly improved by reducing drivers' workload, optimizing resource allocation, alleviating traffic congestion, and minimizing vehicle energy consumption. For AVs, it is essential to accurately measure the pose (translation and rotation) and speed in real time for accurate monitoring, path planning, behavioral decision-making, and control [Urmson and Whittaker, 2008, Guo et al., 2020]. However, the inherent tight connection between AVs and networks makes AVs vulnerable targets of cyberattacks. Therefore, secure pose estimation under attacks has become a crucial problem worth studying.

Vehicle pose estimation is a complex and challenging task that has attracted much attention in recent years. In particular, for a small unmanned aerial vehicle (UAV), a 3D local pose estimation system has been presented in Strohmeier et al. (2018), where the system is realized by fusing 3D position estimations using a loosely coupled extended Kalman filter (EKF) architecture. The data comes from an ultra-wideband transceiver network, an inertial measurement unit sensor, and a barometric pressure sensor. Pose estimation with state or measurement constraints is frequently performed in AV navigation. In view of the inherent constraints, a formulation based on the dynamic potential field has been proposed in Jiang et al. (2020) to express states, measurements, and constraints on connected Riemannian manifolds, after which an information fusion scheme for a dynamic potential field system based on multi-sensor measurements and constraints is designed. It is worth noting that in recent years, due to the fusion of multiple sensors, estimation results are more vulnerable to frequent attacks. Liu

Multimodal Perception and Secure State Estimation for Robotic Mobility Platforms, First Edition.
Xinghua Liu, Rui Jiang, Badong Chen, and Shuzhi Sam Ge.

et al. (2019a) discussed the AV secure pose estimation problem under cyberattacks to deal with possible sensor attacks, and an EKF reconfiguration scheme has been designed to mitigate the influence of sensor attacks.

In the existing research, sensor attacks mainly include denial of service (DoS) attacks [Shi et al., 2021] and deception attacks [Ding et al., 2018b]. DoS attacks are a common attack method used by hackers, who try to make the target machine stop providing services. Deception attacks mean the attacker can rearrange the data in the system to make the sensor or controller receive false data, thus causing the system to fail to function normally. By using a set of random variables in a Bernoulli distribution to describe random deception attacks, a coupled unscented Kalman filter (UKF) has been proposed [Meng and Li, 2018] to propagate the sigma points of the UKF by introducing the coupled terms, and the recursive filtering problem of a class of complex discrete-time networks with random deception attacks has been studied. In Ju et al. (2020), the position sensor deception attack detection and estimation problem is investigated for a local vehicle in a vehicle platoon. A linearized model has been presented to describe the longitudinal dynamics of a local vehicle. In Shi et al. (2021), it is proposed that the attacker behavior is limited only by the frequency and duration of DoS attacks. If the communication links used by the sensor to receive neighbor information lose packets due to DoS attacks, the sensor will give up location estimation. In this chapter, we further assume that the sensor is subject to random deception attacks with a given probability.

This chapter focuses on modeling the AV pose estimation problem with attacks and secure estimation of vehicle poses in a 2D plane. Distinguished from the conventional KF, an UKF-inspired secure recursive estimator is designed to provide estimation, allowing for possible attacks on sensors. By solving several matrix difference equations, the upper bound of the estimation error covariance is guaranteed and correctly updated during the recursive process. The highlights are summarized as follows:

1. The modeling of the system takes into account the occurrence of random deception attacks such that the secure dynamic pose estimation problem has been formulated for AVs.
2. The proposed unscented Kalman-type secure recursive estimator provides a theoretically proven upper bound for error covariance matrices with stable and efficient state estimation.
3. The feasibility and effectiveness of the proposed approach are verified in both a simulated model and the practical AV system, where single and multiple attacks have been considered in the experimental design.

Notation: The following notations are used throughout this chapter. We use $\mathbb{R}^n$ to denote the n-dimensional Euclidean space, and $\mathbb{R}^{m\times n}$ represents the set of all

$m \times n$ matrices. $\mathbb{E}$ denotes the mathematical expectation operator of an underlying probability space, which will be clear from the context. $\mathbf{A} > \mathbf{B}$ implies that both $\mathbf{A}$ and $\mathbf{B}$ are symmetric and $\mathbf{A} - \mathbf{B}$ is positive-definite. We let I be the identity matrix with proper dimensions. Let $||\mathbf{x}||$ and $||\mathbf{A}||$ be the Euclidean norm of a vector $\mathbf{x}$ and a matrix $\mathbf{A}$, respectively. The superscripts $^{\top}$ and $^{-1}$ denote matrix transposition and matrix inverse, respectively.

The remainder of this chapter is organized as follows. Section 7.2 summarizes related work in secure state estimation in cyber-physical systems. Section 7.3 presents the system model and attack model for ground AV pose estimation problem. The estimator design and mathematical proof are presented in Section 7.4; then, Section 7.5 shows simulation results for an illustrative single-input and single-output (SISO) system. Experimental validation and results are shown in Section 7.6. Finally, Section 7.7 concludes the chapter.

7.2 Related Work

A cyber-physical system (CPS) is a complex system with integrated computing, networking, and physical environments. As the interaction between physical and network systems increases, CPSs become more vulnerable to network attacks. Some achievements have been made in secure dynamic state estimation under sensor attacks [Shoukry et al., 2017a, Fawzi et al., 2011]. In Fawzi et al. (2011), the state estimation problem of a linear dynamic system is considered when the measurement data from some sensors is damaged by attackers. In Mishra et al. (2017), when the unknown subset of the sensor is arbitrarily destroyed by the enemy, a secure state estimation algorithm is proposed, and the upper bound of the reachable state estimation error is given.

CPSs play an important role in many fields. In intelligent transportation, regarding AV pose estimation, the relative pose of AVs when driving in a highly dynamic and possibly chaotic environment was studied in Liu et al. (2018b), where a relative pose estimation algorithm based on multiple non-overlapping cameras is proposed; the algorithms are robust even when the number of outliers is overwhelming. In Ding et al. (2020), an enabling multi-sensor fusion-based longitudinal vehicle speed estimator was proposed for four-wheel-independently actuated electric vehicles using a global positioning system and Beidou navigation positioning (GPS-BD) module and a low-cost inertial measurement unit (IMU). Liu et al. (2020) presented a comprehensive evaluation of state-of-the-art sideslip angle estimation methods, with the primary goal of quantitatively revealing their strengths and limitations. Wang et al. (2021) focused on providing a load transfer ratio (LTR) evaluation system that adopts an IMU as the signal input. Unfortunately, less attention is paid to the AV secure pose estimation problem.

In our previous work [Liu et al., 2019c], a secure dynamic pose estimation method based on the filter has been proposed to make the vehicle pose resilient to possible sensor attacks. When all sensors on AVs are benign, the proposed estimator is consistent with the conventional Kalman filtering. On this basis, a vehicle pose estimation based on an UKF under sensor attacks is proposed in this chapter. Compared with other estimators, the proposed estimator in this chapter still follows the framework of the KF, but the next state prediction becomes the expansion and non-linear mapping of the sigma point set. This method has two advantages: (i) the possible complex operation during Jacobian matrix computation for the non-linear process equation can be avoided; and (ii) the approach has better generality in advanced non-linear systems, including those without explicit Jacobian formulation.

In this chapter, we consider the impact of randomly occurring deception attacks (possible sensor attacks) in the design of a secure dynamic pose estimator for AVs. By utilizing the UKF algorithm combined with matrix inequality techniques, we introduce a secure recursive estimation algorithm and derive an upper bound of estimation error covariance by selected optimal estimator parameters. Moreover, the proposed approach can be implemented efficiently in real time and is suitable for recursive computation in applications with limited computational capability.

7.3 Pose Estimation Problem for Ground Vehicles under Attack

In this section, we present the process model, measurement model, and attack model such that the ground vehicle pose estimation problem can be formulated. Although the problem has been modeled in our previous work [Liu et al., 2019c], we still formulate it here for completeness and readers' convenience.

7.3.1 System Model

Consider the following discrete state space model for generality

$$\mathbf{x}_{k+1} = \mathbf{f}(\mathbf{x}_k) + \mathbf{h}(\mathbf{u}_k) + \mathbf{w}_k \tag{7.1}$$

$$\mathbf{z}_k = \mathbf{g}(\mathbf{x}_k) + \mathbf{v}_k \tag{7.2}$$

where k denotes the time index; $\mathbf{f}(\cdot)$ and $\mathbf{g}(\cdot)$ are non-linear process and measurement functions, respectively; and $\mathbf{h}(\mathbf{u}_k)$ is a stochastic function satisfying $\mathbb{E}\{\mathbf{h}(\mathbf{u}_k)|\mathbf{x}_k\} = 0$ for all $\mathbf{x}_k$. $\mathbf{w}_k \sim \mathcal{N}(\mathbf{0}, \mathbf{Q})$ and $\mathbf{v} \sim \mathcal{N}(\mathbf{0}, \mathbf{R})$ denote independent and identically distributed (i.i.d.) Gaussian process and measurement noises with zero mean and covariance matrices $\mathbf{Q} > 0$ and $\mathbf{R} > 0$, respectively.

Two 3D reference frames are used in system modeling: the global frame and the local frame. The global frame (sometimes called the world frame) plays the role of a map on which the vehicle needs to be localized; the local frame (or body frame) moves along the vehicle and is usually the reference for local sensors such as the wheel encoder and IMU. The pose estimation problem aims to estimate the translation and rotation of the local frame with respect to the global frame. As we focus on ground vehicles, projections from 3D to 2D can be applied to reduce the complexity of the model by following certain assumptions [Kelly, 2013]. In particular, the states are defined as

$$\mathbf{x} = \begin{bmatrix} x & y & v & \psi & \dot{\beta} \end{bmatrix}^{\mathrm{T}} \tag{7.3}$$

where x, y and ψ are the coordinates of the vehicle position and the heading on the global $x - y$ plane; v denotes the projection of the vehicle translational velocity onto the local y axis; and $\dot{\beta}$ represents the rotational velocity with respect to the local z axis. In other words, v and $\dot{\beta}$ indicate the forward and rotating velocities that correspond to the vehicle's two manipulating modes: throttle and steering. We further define the control input $\mathbf{u} = [u_v, u_{\dot{\beta}}]^{\mathrm{T}}$ that feeds the throttle and steering into the system motion model.

By incorporating the previous state definition into the vehicle's motion model Eq. (7.1), we have

$$\begin{bmatrix} x \\ y \\ v \\ \psi \\ \dot{\beta} \end{bmatrix}_{k+1} = \begin{bmatrix} x - v(\Delta t \sin \psi) \\ y + v(\Delta t \cos \psi) \\ v + u_v \Delta t \\ \psi + \dot{\beta} \Delta t \\ \dot{\beta} + u_{\dot{\beta}} \Delta t \end{bmatrix}_k + \mathbf{w}_k. \tag{7.4}$$

For the specific formulation of the measurement equation (7.2), we consider a common configuration of sensors [Kelly, 2013] that measure the (translational and rotational) pose x, y, ψ, forward velocity v, and steering angle α as follows

$$\mathbf{z}_k = \begin{bmatrix} z_x \\ z_y \\ z_v \\ z_\psi \\ z_\alpha \end{bmatrix}_k = \begin{bmatrix} x \\ y \\ v \\ \psi \\ \tan^{-1}\left(\frac{L\dot{\beta}}{v}\right) \end{bmatrix}_k + \mathbf{v}_k \tag{7.5}$$

where L denotes the wheelbase between the front and rear wheels of the vehicle. The measurement can be obtained by combining global pose estimation sensors such as satellite navigation systems, visual odometry, and attitude and heading reference systems (AHRSs), and local sensors including wheel encoders and steering angle sensors.

In circumstance where a linear approximation of the measurement equation is required, the Jacobian matrix $\mathbf{G}_k = \frac{\partial \mathbf{g}(\mathbf{x})}{\partial \mathbf{x}}|_{\mathbf{x}=\mathbf{x}_k}$, which needs to be computed at each iteration, can be used for linearization:

$$\mathbf{g}(\mathbf{x}_k) \approx \mathbf{g}(\mathbf{x}_0) + \mathbf{G}_k(\mathbf{x}_k - \mathbf{x}_0). \tag{7.6}$$

Note that only the measurement of α is non-linear, and α is mostly zero with small fluctuations. By selecting $\mathbf{x}_0 = \mathbf{0}$ as the point of interest where $\mathbf{g}(\mathbf{x}_0) = 0$, we have the approximated linear time-varying form of the measurement equation as

$$\mathbf{z}_k = \mathbf{G}_k\mathbf{x}_k + \mathbf{v}_k. \tag{7.7}$$

7.3.2 Attack Model

In this chapter, we assume that the sensors are subject to randomly occurring deception attacks with a given probability. The attack model is described as follows

$$\tilde{\mathbf{z}}_k = \mathbf{z}_k + \gamma_k\mathbf{a}_k = \mathbf{G}_k\mathbf{x}_k + \mathbf{v}_k + \gamma_k\mathbf{a}_k \tag{7.8}$$

where $\tilde{\mathbf{z}}_k$ denotes the measurement with possible attacks, $\mathbf{a}_k$ denotes the information sent by the attacks, and γ_k is a stochastic variable.

Before giving the deception attack model, we make some further assumptions about system knowledge the adversary possesses in order to implement a successful attack. In this chapter, it is assumed that the adversary has sufficient resources and adequate knowledge to arrange a successful attack $\mathbf{a}_k$.

The information $\mathbf{a}_k$ caused by deception attacks can be regarded as $\mathbf{a}_k = -\mathbf{z}_k + \xi_k$, where the nonzero ξ_k satisfying $||\xi_k|| \leq \delta$ is an arbitrary energy-bounded signal. The stochastic variable γ_k is a Bernoulli-distributed white sequence taking values on $\{0, 1\}$ with probabilities

$$\Pr\{\gamma_k = 0\} = \bar{\gamma},\ \Pr\{\gamma_k = 1\} = 1 - \bar{\gamma}.$$

where $\bar{\gamma} \in (0, 1]$ is a known constant. More detailed explanations can be found in Shen et al. (2020).

Remark 7.1 The attack model has the ability to describe the randomly occurring deception attacks; that is, the stochastic variable γ_k is utilized to govern the random nature of sensor attacks on AVs. The false data sent by deception attackers could be identified using algorithms and hardware and software tools. According to the definition of frequentist probability, we may deduce the value of $\bar{\gamma}$ in applications. Hence, the given Bernoulli distribution can properly reveal the random nature of deception attacks.

To derive the main result of this chapter, we will employ the following lemma.

Lemma 7.1 *[Chen and Zheng, 2007b] For any dimension-compatible matrices* **D**, **E**, *and a scalar* $\varepsilon > 0$, *the following inequality holds:*

$$\mathbf{D}\mathbf{E} + \mathbf{E}^{\top}\mathbf{D}^{\top} \leq \varepsilon\mathbf{D}\mathbf{D}^{\top} + \varepsilon^{-1}\mathbf{E}\mathbf{E}^{\top}.$$

7.4 Design of the Unscented Kalman Filter

The UKF uses unscented transformation (UT) to represent a random variable by using a number of deterministically selected sample points (called sigma points). These points capture the mean and covariance of the random variable and, when propagated through the true non-linear system, capture the posterior mean and covariance accurately.

Denote the one-step prediction error and the estimation error as $\mathbf{e}_{k+1|k} = \mathbf{x}_{k+1} - \hat{\mathbf{x}}_{k+1|k}$ and $\mathbf{e}_k = \mathbf{x}_k - \hat{\mathbf{x}}_k$, respectively. The one-step prediction error covariance matrix $\mathbf{P}_{k+1|k}$ and the estimation error covariance matrix $\mathbf{P}_{k+1}$ can be obtained as follows:

$$\mathbf{P}_{k+1|k} = \mathbb{E}\{[\mathbf{x}_{k+1} - \hat{\mathbf{x}}_{k+1|k}][\mathbf{x}_{k+1} - \hat{\mathbf{x}}_{k+1|k}]^{\top}\}, \tag{7.9a}$$

$$\mathbf{P}_{k+1} = \mathbb{E}\{[\mathbf{x}_{k+1} - \hat{\mathbf{x}}_{k+1}][\mathbf{x}_{k+1} - \hat{\mathbf{x}}_{k+1}]^{\top}\}. \tag{7.9b}$$

We are now ready to conduct the one-step prediction error matrix in terms of the solvability of recursive Riccati difference equations and obtain the parameter gain matrix of the UKF, which is developed in the following theorem.

Theorem 7.1 *Consider the discrete kinematic equation* (7.1) *suffering from attacks as* (7.8). *For any given positive constants* $\varepsilon_k, k = 0, 1, 2, \cdots$ *and the initial condition* $\mathbf{x}_0, \hat{\mathbf{x}}_0 = \mathbb{E}\{\mathbf{x}_0\}, \mathbf{\Pi}_0 = \mathbf{P}_0, \mathbf{\Sigma}_0 = \mathbb{E}\{\mathbf{x}_0\mathbf{x}_0^{\top}\}$, *we can derive that the parameter gain matrix of the UKF is given as follows*

$$\begin{aligned}\mathbf{K}_{k+1} = {} & \alpha_4\mathbf{P}_{k+1|k}\mathbf{G}_k^{\top}[\alpha_1\mathbf{G}_k\mathbf{P}_{k+1|k}\mathbf{G}_k^{\top} + \alpha_2\mathbf{G}_k\mathbf{\Sigma}_{k+1}\mathbf{G}_k^{\top} \\ & + \alpha_3\mathbf{I} + \mathbf{R}]^{-1},\end{aligned} \tag{7.10}$$

where $\alpha_1 = [1 + (1 - \bar{\gamma})\varepsilon_k]\bar{\gamma}^2$, $\alpha_2 = \bar{\gamma}(1 - \bar{\gamma})(1 + \varepsilon_k)$, $\alpha_3 = (1 - \bar{\gamma}^2)\varepsilon_k^{-1}\delta^2 + (1 - \bar{\gamma})\delta^2$, *and* $\alpha_4 = \bar{\gamma} + (\bar{\gamma} - \bar{\gamma}^2)\varepsilon_k$. *The upper bound for the estimation error covariance is* $\mathbf{\Pi}_{k+1}$, *which can be recursively calculated by* (7.21).

Proof: Step 1: Initialization.

To calculate the statistics of a random variable that undergoes a non-linear transformation, a matrix χ is generated using $2n + 1$ weighted **sigma points**. The computation algorithm begins with the initial conditions

$$\hat{\mathbf{x}}_0 = \mathbb{E}\{\mathbf{x}_0\}, \ \mathbf{P}_0 = \mathbb{E}\left\{(\mathbf{x}_0 - \hat{\mathbf{x}}_0)(\mathbf{x}_0 - \hat{\mathbf{x}}_0)^{\top}\right\}. \tag{7.11}$$

Step 2: Generation of sigma points.
We calculate the UT sampling as follows

$$\begin{cases} \chi_{i,k|k} = \hat{\mathbf{x}}_{k|k}, \ i = 0 \\ \chi_{i,k|k} = \hat{\mathbf{x}}_{k|k} + \left(\sqrt{(n+\lambda)\mathbf{P}_{k|k}}\right)_i, \ i = 1, 2, ..., n \\ \chi_{i,k|k} = \hat{\mathbf{x}}_{k|k} - \left(\sqrt{(n+\lambda)\mathbf{P}_{k|k}}\right)_i, \ i = n+1, ..., 2n \end{cases} \tag{7.12}$$

$$\begin{cases} \omega_i^{(m)} = \dfrac{\lambda}{n+\lambda}, \ i = 0 \\ \omega_i^{(c)} = \dfrac{\lambda}{n+\lambda} + (1 - \alpha^2 + \beta), \ i = 0 \\ \omega_i^{(m)} = \omega_i^{(c)} = \dfrac{1}{2(n+\lambda)}, \ i = 1, 2, ..., 2n \end{cases} \tag{7.13}$$

where $\lambda = \alpha^2(n+\kappa) - n$, α is the proportion factor, and the distribution distance of the particles can be adjusted by changing the value of α to reduce the error. Parameters κ and β can be tuned and are generally set to 0 and 2, respectively. $(\sqrt{(n+\lambda)\mathbf{P}_{k|k}})_i$ is the ith column of the square root of the matrix, $\omega_i^{(m)}$ is the weighted mean, and $\omega_i^{(c)}$ is the weighted covariance.

Step 3: A one-step prediction is made for sigma sampling points to get the state prediction value and prediction covariance of each particle. First, we calculate the state prediction value as follows

$$\chi_{i,k+1|k} = \mathbf{f}\left(\chi_{i,k|k}\right), \tag{7.14a}$$

$$\hat{\mathbf{x}}_{k+1|k} = \sum_{i=0}^{2n} \omega_i^{(m)} \chi_{i,k+1|k}.. \tag{7.14b}$$

And from (7.9a), we know that

$$\mathbf{P}_{k+1|k} = \sum_{i=0}^{2n} \omega_i^{(c)} [\chi_{i,k+1|k} - \hat{\mathbf{x}}_{k+1|k}][\chi_{i,k+1|k} - \hat{\mathbf{x}}_{k+1|k}]^{\mathrm{T}} + \mathbf{Q}. \tag{7.15}$$

Then we have

$$\begin{aligned} \hat{\mathbf{x}}_{k+1} &= \hat{\mathbf{x}}_{k+1|k} + \mathbf{K}_{k+1}(\tilde{\mathbf{z}}_{k+1} - \bar{\gamma}\mathbf{G}_k\hat{\mathbf{x}}_{k+1|k}) \\ &= \hat{\mathbf{x}}_{k+1|k} + \mathbf{K}_{k+1}(\mathbf{G}_k\mathbf{x}_{k+1} + \mathbf{v}_{k+1} + \gamma_{k+1}\mathbf{a}_{k+1} \\ &\quad - \bar{\gamma}\mathbf{G}_k\hat{\mathbf{x}}_{k+1|k}). \end{aligned} \tag{7.16}$$

Step 4: Posterior error.

Since $\mathbf{e}_k = \mathbf{x}_k - \hat{\mathbf{x}}_k$, and if we plug in $\mathbf{a}_{k+1} = -\mathbf{G}_k\mathbf{x}_{k+1} - \mathbf{v}_{k+1} + \xi_{k+1}$, it can be obtained that

$$\mathbf{e}_{k+1} = \mathbf{x}_{k+1} - \hat{\mathbf{x}}_{k+1}$$

$$
\begin{aligned}
&= \mathbf{x}_{k+1} - \hat{\mathbf{x}}_{k+1|k} - \mathbf{K}_{k+1}(\mathbf{G}_k\mathbf{x}_{k+1} + \mathbf{v}_{k+1} \\
&\quad + \gamma_{k+1}\mathbf{a}_{k+1} - \bar{\gamma}\mathbf{G}_k\hat{\mathbf{x}}_{k+1|k}) \\
&= (\mathbf{I} - \bar{\gamma}\mathbf{K}_{k+1}\mathbf{G}_k)\mathbf{e}_{k+1|k} - \mathbf{K}_{k+1}(\gamma_{k+1}\mathbf{a}_{k+1} \\
&\quad + \mathbf{v}_{k+1} + (1-\bar{\gamma})\mathbf{G}_k\mathbf{x}_{k+1}) \\
&= (\mathbf{I} - \bar{\gamma}\mathbf{K}_{k+1}\mathbf{G}_k)\mathbf{e}_{k+1|k} - \mathbf{K}_{k+1}(-\gamma_{k+1} \\
&\quad \times \mathbf{G}_k\mathbf{x}_{k+1} - \gamma_{k+1}\mathbf{v}_{k+1} + \gamma_{k+1}\xi_{k+1} \\
&\quad + \mathbf{v}_{k+1} + (1-\bar{\gamma})\mathbf{G}_k\mathbf{x}_{k+1}).
\end{aligned} \tag{7.17}
$$

Step 5: Posterior covariance.
From (7.9b), we know that

$$
\begin{aligned}
\mathbf{P}_{k+1} &= \mathbb{E}\{\mathbf{e}_{k+1}\mathbf{e}_{k+1}^{\top}\} \\
&= \mathbb{E}\{[(\mathbf{I} - \bar{\gamma}\mathbf{K}_{k+1}\mathbf{G}_k)\mathbf{e}_{k+1|k} - \mathbf{K}_{k+1}(-\gamma_{k+1}\mathbf{G}_k\mathbf{x}_{k+1} \\
&\quad - \gamma_{k+1}\mathbf{v}_{k+1} + \gamma_{k+1}\xi_{k+1} + \mathbf{v}_{k+1} + (1-\bar{\gamma}) \\
&\quad \times \mathbf{G}_k\mathbf{x}_{k+1})][(\mathbf{I} - \bar{\gamma}\mathbf{K}_{k+1}\mathbf{G}_k)\mathbf{e}_{k+1|k} - \mathbf{K}_{k+1} \\
&\quad \times (-\gamma_{k+1}\mathbf{G}_k\mathbf{x}_{k+1} - \gamma_{k+1}\mathbf{v}_{k+1} + \gamma_{k+1}\xi_{k+1} \\
&\quad + \mathbf{v}_{k+1} + (1-\bar{\gamma})\mathbf{G}_k\mathbf{x}_{k+1})]^{\top}\}.
\end{aligned} \tag{7.18}
$$

Then we can obtain that

$$
\begin{aligned}
\mathbf{P}_{k+1} &= (\mathbf{I} - \bar{\gamma}\mathbf{K}_{k+1}\mathbf{G}_k)\mathbf{P}_{k+1|k}(\mathbf{I} - \bar{\gamma}\mathbf{K}_{k+1}\mathbf{G}_k)^{\top} \\
&\quad - (1-\bar{\gamma})(\mathbf{I} - \bar{\gamma}\mathbf{K}_{k+1}\mathbf{G}_k)\mathbb{E}\{\mathbf{e}_{k+1|k}\xi_{k+1}^{\top}\}\mathbf{K}_{k+1}^{\top} \\
&\quad - (1-\bar{\gamma})\mathbf{K}_{k+1}\mathbb{E}\{\xi_{k+1}\mathbf{e}_{k+1|k}^{\top}\}(\mathbf{I} - \bar{\gamma}\mathbf{K}_{k+1}\mathbf{G}_k)^{\top} \\
&\quad - \bar{\gamma}(1-\bar{\gamma})\mathbf{K}_{k+1}\mathbf{G}_k\mathbb{E}\{\mathbf{x}_{k+1}\xi_{k+1}^{\top}\}\mathbf{K}_{k+1}^{\top} \\
&\quad - \bar{\gamma}(1-\bar{\gamma})\mathbf{K}_{k+1}\mathbb{E}\{\xi_{k+1}\mathbf{x}_{k+1}^{\top}\}\mathbf{G}_k^{T}\mathbf{K}_{k+1}^{\top} \\
&\quad + \bar{\gamma}(1-\bar{\gamma})\mathbf{K}_{k+1}\mathbf{G}_k\mathbb{E}\{\mathbf{x}_{k+1}\mathbf{x}_{k+1}^{\top}\}\mathbf{G}_k^{\top}\mathbf{K}_{k+1}^{\top} \\
&\quad + (1-\bar{\gamma})\mathbf{K}_{k+1}\xi_{k+1}\xi_{k+1}^{\top}\mathbf{K}_{k+1}^{\top} \\
&\quad + \bar{\gamma}\mathbf{K}_{k+1}\mathbf{R}\mathbf{K}_{k+1}^{\top}.
\end{aligned} \tag{7.19}
$$

By **Lemma** 7.1 and applying the property of matrix trace, we have that

$$
\begin{aligned}
\mathbf{P}_{k+1} &\leq (\mathbf{I} - \bar{\gamma}\mathbf{K}_{k+1}\mathbf{G}_k)\mathbf{P}_{k+1|k}(\mathbf{I} - \bar{\gamma}\mathbf{K}_{k+1}\mathbf{G}_k)^{\top} \\
&\quad - (1-\bar{\gamma})[\varepsilon_k(\mathbf{I} - \bar{\gamma}\mathbf{K}_{k+1}\mathbf{G}_k)\mathbb{E}(\mathbf{e}_{k+1}\mathbf{e}_{k+1}^{T}) \\
&\quad \times (\mathbf{I} - \bar{\gamma}\mathbf{K}_{k+1}\mathbf{G}_k)^{T} + \varepsilon_k^{-1}\delta^2\mathbf{K}_{k+1}\mathbf{K}_{k+1}^{\top}] \\
&\quad - \bar{\gamma}(1-\bar{\gamma})\mathbf{K}_{k+1}[\varepsilon_k\mathbf{G}_k\boldsymbol{\Sigma}_{k+1}\mathbf{G}_k^{T} + \varepsilon_k^{-1}\delta^2] \\
&\quad \times \mathbf{K}_{k+1}^{\top} + \bar{\gamma}(1-\bar{\gamma})\mathbf{K}_{k+1}\mathbf{G}_k\boldsymbol{\Sigma}_{k+1}\mathbf{G}_k^{T}\mathbf{K}_{k+1}^{\top} \\
&\quad + (1-\bar{\gamma})\mathbf{K}_{k+1}\delta^2\mathbf{K}_{k+1}^{\top} + \bar{\gamma}\mathbf{K}_{k+1}\mathbf{R}\mathbf{K}_{k+1}^{\top},
\end{aligned}
$$

$$
\begin{aligned}
&=[1+(1-\bar{\gamma})\varepsilon_k](\mathbf{I}-\bar{\gamma}\mathbf{K}_{k+1}\mathbf{G}_k)\mathbf{P}_{k+1|k}(\mathbf{I}-\bar{\gamma}\mathbf{K}_{k+1}\mathbf{G}_k)^{\mathrm{T}} \\
&\quad+[\bar{\gamma}(1-\bar{\gamma})(1+\varepsilon_k)]\mathbf{K}_{k+1}\mathbf{G}_k\mathbf{\Sigma}_{k+1}\mathbf{G}_k^{\mathrm{T}}\mathbf{K}_{k+1}^{\mathrm{T}} \\
&\quad+[(1-\bar{\gamma}^2)\varepsilon_k^{-1}\delta^2+(1-\bar{\gamma})\delta^2]\mathbf{K}_{k+1}\mathbf{K}_{k+1}^{\mathrm{T}} \\
&\quad+\bar{\gamma}\mathbf{K}_{k+1}\mathbf{R}\mathbf{K}_{k+1}^{\mathrm{T}}
\end{aligned}
\tag{7.20}
$$

where

$$
\begin{aligned}
\mathbf{\Sigma}_{k+1} &=\mathbb{E}\{\mathbf{x}_{k+1}\mathbf{x}_{k+1}^{\mathrm{T}}\} \\
&=\mathbb{E}\{\left(\mathbf{f}(\mathbf{x}_k)+\mathbf{h}(\mathbf{u}_k)+\mathbf{w}_k\right)\left(\mathbf{f}(\mathbf{x}_k)+\mathbf{h}(\mathbf{u}_k)+\mathbf{w}_k\right)^{\mathrm{T}}\} \\
&=\sum_{i=0}^{2n}\omega_i^{(c)}(\chi_{i,k+1|k}\cdot\chi_{i,k+1|k}^{\mathrm{T}})+\mathbf{h}(\mathbf{u}_k)\mathbf{h}(\mathbf{u}_k)^{\mathrm{T}}+\mathbf{Q}.
\end{aligned}
$$

Define

$$
\begin{aligned}
\mathbf{\Pi}_{k+1} &= [1+(1-\bar{\gamma})\varepsilon_k](\mathbf{I}-\bar{\gamma}\mathbf{K}_{k+1}\mathbf{G}_k)\mathbf{P}_{k+1|k}(\mathbf{I}-\bar{\gamma}\mathbf{K}_{k+1}\mathbf{G}_k)^{\mathrm{T}} \\
&\quad+[\bar{\gamma}(1-\bar{\gamma})(1+\varepsilon_k)]\mathbf{K}_{k+1}\mathbf{G}_k\mathbf{\Sigma}_{k+1}\mathbf{G}_k^{\mathrm{T}}\mathbf{K}_{k+1}^{\mathrm{T}} \\
&\quad+[(1-\bar{\gamma}^2)\varepsilon_k^{-1}\delta^2+(1-\bar{\gamma})\delta^2]\mathbf{K}_{k+1}\mathbf{K}_{k+1}^{\mathrm{T}} \\
&\quad+\bar{\gamma}\mathbf{K}_{k+1}\mathbf{R}\mathbf{K}_{k+1}^{\mathrm{T}}.
\end{aligned}
\tag{7.21}
$$

Taking the partial derivation of the trace of the matrix $\mathbf{\Pi}_{k+1}$ with respect to $\mathbf{K}_{k+1}$ and letting the derivative be zero, we can obtain that

$$
\begin{aligned}
&-2[\bar{\gamma}+(\bar{\gamma}-\bar{\gamma}^2)\varepsilon_k](\mathbf{I}-\bar{\gamma}\mathbf{K}_{k+1}\mathbf{G}_k)\mathbf{P}_{k+1|k}\mathbf{G}_k^{\mathrm{T}} \\
&+2\bar{\gamma}(1-\bar{\gamma})(1+\varepsilon_k)\mathbf{K}_{k+1}\mathbf{G}_k\mathbf{\Sigma}_{k+1}\mathbf{G}_k^{\mathrm{T}} \\
&+2[(1-\bar{\gamma}^2)\varepsilon_k^{-1}\delta^2+(1-\bar{\gamma})\delta^2]\mathbf{K}_{k+1} \\
&+2\bar{\gamma}\mathbf{K}_{k+1}\mathbf{R}=0.
\end{aligned}
\tag{7.22}
$$

It follows that

$$
\begin{aligned}
&\mathbf{K}_{k+1}[\alpha_1\mathbf{G}_k\mathbf{P}_{k+1|k}\mathbf{G}_k^{\mathrm{T}}+\alpha_2\mathbf{G}_k\mathbf{\Sigma}_{k+1}\mathbf{G}_k^{\mathrm{T}}+\alpha_3\mathbf{I}+\mathbf{R}] \\
&=\alpha_4\mathbf{P}_{k+1|k}\mathbf{G}_k^{\mathrm{T}}.
\end{aligned}
$$

Since $\alpha_1\mathbf{G}_k\mathbf{P}_{k+1|k}\mathbf{G}_k^{\mathrm{T}}+\alpha_2\mathbf{G}_k\mathbf{\Sigma}_{k+1}\mathbf{G}_k^{\mathrm{T}}+\alpha_3\mathbf{I}+\mathbf{R}$ is a positive definite matrix, we know (7.10) holds, and the proof is complete. □

As we can see in (7.10) and (7.15), we need $\mathcal{O}(n^3)$ operations to compute the Kalman gain and the covariance matrix $P_{k+1|k}$, where $n=5$ in this chapter. This indicates that the secure recursive estimator can be treated in a short time without a high-performance computer.

Remark 7.2 From (7.21), it can be seen that the larger δ leads to a bigger upper bound of the estimation error covariance, which means the estimation performance deteriorates with increased δ.

Remark 7.3 According to the matrix inequality technique of **Lemma** 7.1, we can arbitrarily choose the positive constant ε_k in **Theorem 1** from the theoretical point of view. However, a value of ε_k that is too large or too small may influence the estimation performance. In practice or experimental validation, we select the appropriate positive constant ε_k based on experience to achieve better estimation performance.

How to obtain a secure pose estimation for the discrete kinematic equation (7.1) for AVs in the presence of deception attacks (sensor attacks) has remained an open problem until now. The goal of this chapter is to propose an algorithm that enables us to estimate the pose states of the AV in such a way that:

1. If no sensors are comprised, i.e. $\overline{\gamma} = 1$ and $\gamma_k = 0$ with probability 1 for all k, the estimate coincides with the standard KF.
2. If fewer than half of the pose states are compromised by randomly occurring deception attacks, it still gives a stable estimate of the pose states: i.e. an upper bound for the estimation error covariance is guaranteed.

According to **Theorem 1**, the calculation framework can be summarized as the following **Algorithm 1**.

7.5 Numeric Simulation

We first run the proposed approach based on a simple but effective single-input and single-output (SISO) system. Consider

$$x_{k+1} = \frac{10000}{x_k} + u_k + w_k \tag{7.23}$$

$$z_k = x_k + v_k \tag{7.24}$$

where the control $u_k = 1$ and the attack $a_k = 30$. The attack-related parameters are set as $\delta = 30$ and $\epsilon_k = 0.001$. The process and measurement noise levels are $Q = 1$ and $R = 0.01$, respectively. The UKF parameters are $\alpha = 1e^{-3}$, $\kappa = 0$, and $\beta = 2$. The system is initialized as $x_0 = 50$.

We select this non-linear process model and the identical measurement model because the proposed approach does not apply to non-linear measurements directly. In other words, a linear approximation of the measurement equation is always required, similar to Liu et al. (2019c). The identical measurement equation allows direct comparison between the proposed approach and Liu et al. (2019c) and avoids unnecessary bias from measurement equation linearization. Although a simple model is used in the simulation, the experiments in the next section present the algorithm's performance based on a non-linear measurement model.

Algorithm 5 UKF-based secure recursive estimator

Step 1. Initialization:

1. Set the values of the initial pose state $\mathbf{x}_0$, initial estimate state $\hat{\mathbf{x}}_0$, initial estimation error covariance matrix, $\mathbf{P}_0$ and initial state covariance matrix $\boldsymbol{\Sigma}_0$;
2. Set the value of $\bar{\gamma}$, and determine the value of δ (the norm bound of arbitrary signal ξ_k);
3. Set the control input signal $\mathbf{u}_k$, i.e. the translational and rotational accelerations for the AV;
4. Let $\boldsymbol{\Pi}_0 = \mathbf{P}_0$, and choose the proper ε_k for all k to calculate α_1, α_2, α_3, and α_4;
5. Set the discrete time index $k = 0$.

Step 2. State covariance matrix $\boldsymbol{\Sigma}_{k+1}$ is updated as follows:

$$\boldsymbol{\Sigma}_{k+1} = \sum_{i=0}^{2n} \omega_i^{(c)} (\chi_{i,k+1|k} \cdot \chi_{i,k+1|k}^{\mathrm{T}}) + \mathbf{h}(\mathbf{u}_k)\mathbf{h}(\mathbf{u}_k)^{\mathrm{T}} + \mathbf{Q}.$$

Step 3. The secure recursive estimator gain $\mathbf{K}_{k+1}$ and $\boldsymbol{\Pi}_{k+1}$ are calculated as follows:

$$\begin{aligned}
\mathbf{K}_{k+1} &= \alpha_4 \mathbf{P}_{k+1|k}\mathbf{G}_k^{\mathrm{T}}[\alpha_1 \mathbf{G}_k \mathbf{P}_{k+1|k}\mathbf{G}_k^{\mathrm{T}} + \alpha_2 \mathbf{G}_k \boldsymbol{\Sigma}_{k+1}\mathbf{G}_k^{\mathrm{T}} \\
&\quad + \alpha_3 \mathbf{I} + \mathbf{R}]^{-1}, \\
\hat{\mathbf{x}}_{k+1} &= \hat{\mathbf{x}}_{k+1|k} + \mathbf{K}_{k+1}(\tilde{\mathbf{z}}_{k+1} - \bar{\gamma}\mathbf{G}_k\hat{\mathbf{x}}_{k+1|k}) \\
&= \hat{\mathbf{x}}_{k+1|k} + \mathbf{K}_{k+1}(\mathbf{G}_k\mathbf{x}_{k+1} + \mathbf{v}_{k+1} + \gamma_{k+1}\mathbf{a}_{k+1} \\
&\quad - \bar{\gamma}\mathbf{G}_k\hat{\mathbf{x}}_{k+1|k}), \\
\boldsymbol{\Pi}_{k+1} &= [1+(1-\bar{\gamma})\varepsilon_k](\mathbf{I}-\bar{\gamma}\mathbf{K}_{k+1}\mathbf{G}_k)\mathbf{P}_{k+1|k}(\mathbf{I}-\bar{\gamma}\mathbf{K}_{k+1}\mathbf{G}_k)^{\mathrm{T}} \\
&\quad + [\bar{\gamma}(1-\bar{\gamma})(1+\varepsilon_k)]\mathbf{K}_{k+1}\mathbf{G}_k\boldsymbol{\Sigma}_{k+1}\mathbf{G}_k^{\mathrm{T}}\mathbf{K}_{k+1}^{\mathrm{T}} \\
&\quad + [(1-\bar{\gamma}^2)\varepsilon_k^{-1}\delta^2 + (1-\bar{\gamma})\delta^2]\mathbf{K}_{k+1}\mathbf{K}_{k+1}^{\mathrm{T}} \\
&\quad + \bar{\gamma}\mathbf{K}_{k+1}\mathbf{R}\mathbf{K}_{k+1}^{\mathrm{T}}.
\end{aligned}$$

Step 4. Set $k=k+1$, and go to Step 2.

Figure 7.1 shows the simulation performance under different attack intensities. The legends "UKF", "EKF", and "KF" denote the performance of the proposed approach, the method in Liu et al. (2019c), and the conventional KF. We select EKF and KF for comparison because these two methods are classical filters that have been widely used in practice. The comparison to classical methods gives readers a more intuitive illustration of the gain from the proposed approach. The first column illustrates the ground truth state and the measurement with attacks. As $\bar{\gamma}$ increases, the probability of attacks decreases and the measurements are less disturbed by the environmental aspect. The second column shows the estimation error, where EKF and UKF result in similar and stable estimation errors that are less influenced by attack intensity $\bar{\gamma}$. The KF leads to a large estimation error when $\bar{\gamma}$ is small, but the estimation becomes much more accurate when there is a

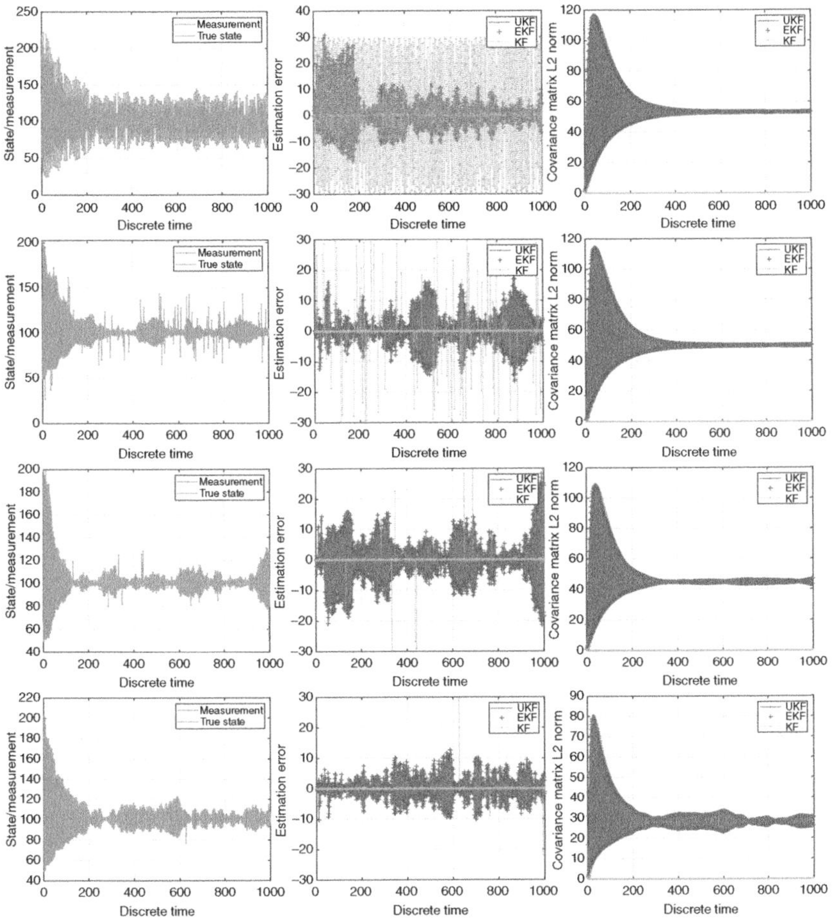

Figure 7.1 Simulation performance of the proposed filter ("UKF"), the filter in Liu et al. (2019c) ("EKF"), and the Kalman filter ("KF"). The first column illustrates the ground truth state and measurement values. The second column shows estimation errors. The third column represents the norms of the estimator covariance matrices $\mathbf{\Pi}$, which are real positive numbers in this case. The first to fourth rows show results where $\overline{\gamma} = 0.1, 0.9$, 0.99, and 0.999, respectively.

small chance of being attacked. However, as the KF does not consider the attack issue, estimation accuracy may deteriorate suddenly at a discrete time around 620 with $\overline{\gamma} = 0.999$. The third column presents the norm of the estimator covariance matrix, which is a number in the SISO system. It is found that the KF gives a completely wrong estimation error covariance matrix by comparing the second and the third columns in Figure 7.1: the KF outputs a nearly zero estimation error covariance matrix, but the estimation errors are quite large under attacks. On the

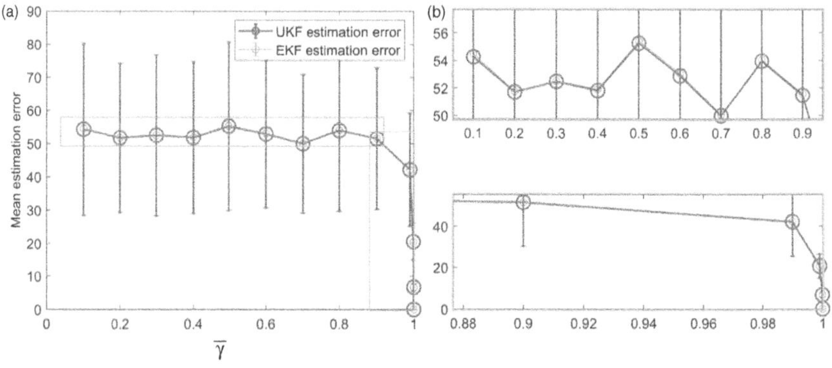

Figure 7.2 The estimation squared error bar graph with respect to $\overline{\gamma}$. (a) The mean and standard deviation of the squared errors under different non-attack probabilities in 500 simulations. (b) Local details of the error graph. "UKF" and "EKF" denote the proposed filter and the approach in Liu et al. (2019c), respectively.

other hand, the proposed approach and Liu et al. (2019c) both provide reliable error upper-bound covariance matrices.

To test stability and robustness under random noises and attacks, we repeat the simulation 500 times and compute the error as $e = \frac{1}{N}\sum_N\{(x-\hat{x})^2\}$ where N denotes the total number of discrete time indexes. The mean and standard deviation of error e with respect to non-attack probabilities are illustrated in Figure 7.2. From the results, it is noted that the UKF performs slightly better than the EKF for the simulated dynamic system. Moreover, both the UKF and EKF estimation errors stay almost unchanged with $\overline{\gamma}$ from 0.1 to 0.9, but the errors drop dramatically with $\overline{\gamma}$ from 0.99 to 1.

7.6 Experiments

We apply the proposed approach to the ground vehicle pose estimation problem formulated in Section 7.3. The following attack signal is added to the measurement:

$$\mathbf{a}_k = \begin{bmatrix}10 & 10 & 0 & 0 & 0\end{bmatrix}^{\mathrm{T}}. \tag{7.25}$$

The control signal reflects common driving behaviors, and details can be found in our previous work [Liu et al., 2019c]. The attack-related parameters are set as $\delta = 14$ and $\epsilon_k = 0.001$. The process and measurement noise levels are

$\mathbf{Q} = \mathrm{diag}(0.001^2, 0.001^2, 0.001^2, 0.0005^2, 0.0001^2)$ and
$\mathbf{R} = \mathrm{diag}(1.0^2, 1.0^2, \mathtt{deg2rad}^2(1), 0.5^2, \mathtt{deg2rad}^2(2))$,

respectively, where deg2rad denotes the conversion from degree to radian. The UKF parameters are $\alpha = 1e^3$, $\kappa = 0$, and $\beta = 2$. The system is initialized as $\mathbf{x}_0 = [0, 0, 0.001, \texttt{deg2rad}(70), 0]^\top$.

In practice, the process and measurement noises of the filtering algorithms are unknown and need to be estimated using modeling or statistical approaches. In this chapter, the process and measurement noises are identical to the ground truth, which is an optimal selection. All other shared parameters are kept the same for all the methods for a fair comparison.

7.6.1 General Performance

The pose estimation errors for selected states under different non-attack probabilities can be found in Figure 7.3. Note that we do not show the EKF performance [Liu et al., 2019c] in this section since the EKF and UKF do not share the same parameters; thus it is hard to compare these two methods fairly with different configurations. From the results, it is noted that the proposed estimation may perform worse when there is less chance of an attack, as shown in the last row of the figure, where $\overline{\gamma} = 0.999$ indicates that the chance of an attack is extremely low. In such a case, the conventional KF performs well. If there are frequent attacks, the proposed estimator generally has more stable results than the KF, with fewer sudden fluctuations. However, unlike the KF, the proposed approach does not guarantee the best linear estimation performance in the minimum mean-square-error sense, since we have only derived an upper bound of the estimation error covariance matrix. In this case, it is not a surprise to have poor estimation accuracy in some states, such as y and ψ.

The diagonal elements in the estimation error covariance matrices with respect to various non-attack probabilities ($\overline{\gamma} = 0.3, 0.7$, and 0.999) are illustrated in Figure 7.4, where we can monitor the estimation quality in different states in real time. The results show that lower upper bounds are derived with larger $\overline{\gamma}$. Note that a larger $\overline{\gamma}$ does not ensure a higher estimation accuracy but only gives a narrower range of estimation errors.

7.6.2 Influence of Parameters

It is observed during the experiments that the parameters of the estimator have a significant influence on performance. There are three configurable parameters in the proposed approach: δ, ϵ, and $\overline{\gamma}$. Theoretically, δ should be set based on the attacks. However, the attack signal is unknown in practice, so δ is set based on our experience and prediction of attacks. A conservative and large δ may lead to a large $\mathbf{\Pi}$, while a small δ may get the violation of inequality (7.20) and invalidate the error covariance matrix $\mathbf{\Pi}$. A large ϵ usually leads to divergence of the estimation;

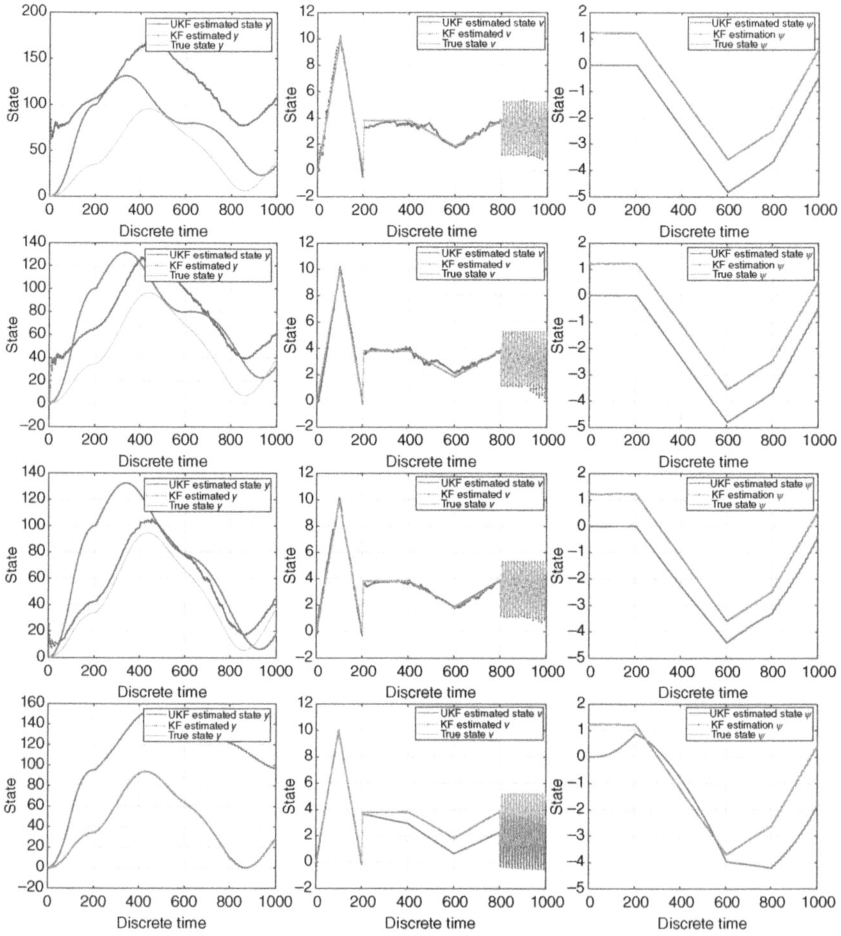

Figure 7.3 Pose estimation of selected states y (first column), v (second column), and ψ (third column) of the proposed filter ("UKF") and the Kalman filter ("KF"). The first to fourth rows show the results where $\overline{\gamma} = 0.3, 0.7, 0.9$, and 0.999, respectively.

thus, it is set to a small value in all experiments. Finally, $\overline{\gamma}$ is set as the non-attack probability of the attack signal. Practically, $\overline{\gamma}$ is unknown and can be configured according to the attack threat level.

In addition, the UKF parameters influence the algorithm's performance. An appropriate selection of α, κ, and β is required to adjust the distribution of sigma points for the dynamic system. Still, tuning is necessary during the experiments since there is no direct guidance for UKF parameter selection.

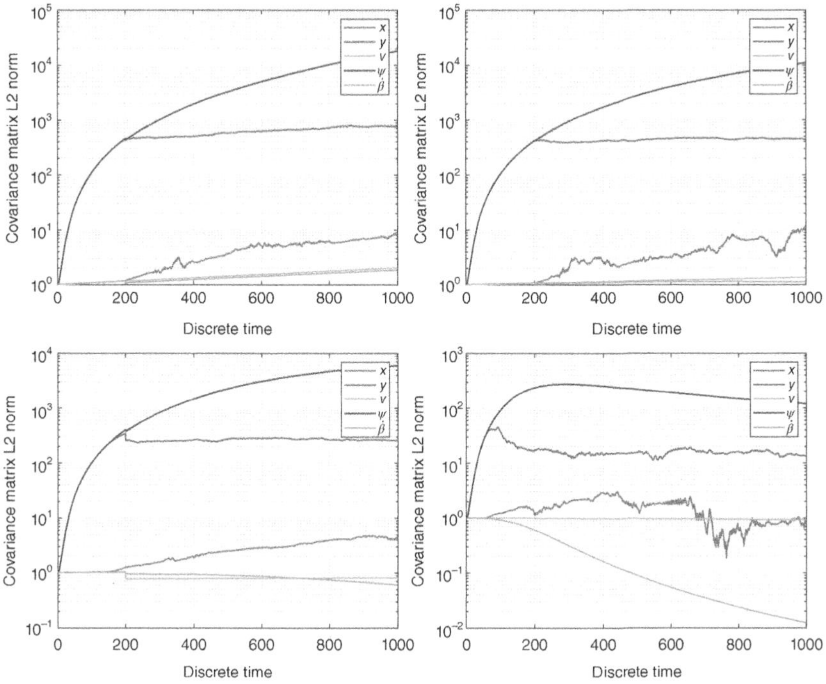

Figure 7.4 The diagonal elements in the estimation covariance matrices **Π**. The first to fourth rows show results where $\overline{\gamma} = 0.3, 0.7, 0.9$, and 0.999, respectively.

7.7 Conclusion

In this chapter, a recursive pose estimator inspired by the unscented Kalman filter has been designed to tackle the secure vehicle pose estimation problem under random deception attacks. The estimator minimizes the upper bound of the estimation error covariance, which can be solved very efficiently in real time and is suitable for recursive computation in online applications. Simulations and experiments have been designed to validate the effectiveness of the proposed estimation approach.

8

Secure Dynamic State Estimation with a Decomposing Kalman Filter

8.1 Introduction

In modern societies, critical infrastructures are mostly managed by cyber-physical systems (CPSs): systems that embed widespread sensing, networking, computation, and control into physical spaces. Examples of CPSs include smart grids, autonomous automobile systems, process control systems, robotics systems, and so on [Khaitan and McCalley, 2015]. A wide variety of motivations exist for launching an attack on CPSs, ranging from financial reasons, e.g. reducing the electricity bill, all the way to terrorism, e.g. threatening the life of an entire population by controlling electricity and other life-critical resources. Any attack on safety-critical CPSs may significantly hamper the economy and even lead to the loss of human life. For instance, the Stuxnet virus, which attacked industrial supervisory control and data acquisition (SCADA) systems, was detected in July 2010 [Chen, 2010, Fidler, 2011]. Recently, the Ukraine power plant hack provides a clear example of the catastrophic outcomes of successful attacks on SCADA systems. Since several incidents have illustrated the susceptibility of CPSs to attacks, the security problem has received much attention, and securing CPSs against malicious attacks or communication failures has become an important issue [Pasqualetti et al., 2015]. Hence, the research community has acknowledged the importance of security in CPSs and has focused on designing secure detection, estimation, and control strategies [Cárdenas et al., 2008].

Significant efforts have been invested into secure dynamic state estimation against sensor attacks in the past few years. Liu et al. (2009) illustrate how an adversary can inject a stealthy input into the measurements to change the state estimation without being detected by the bad data detector. Sandberg et al. (2010b) consider how to find a sparse stealthy input that enables the adversary to launch an attack with a minimum number of compromised sensors. Xie et al. (2011) further illustrate that stealthy integrity attacks on state estimation can lead to a financial gain in the electricity market for the adversary. In addition,

Multimodal Perception and Secure State Estimation for Robotic Mobility Platforms, First Edition.
Xinghua Liu, Rui Jiang, Badong Chen, and Shuzhi Sam Ge.

robustification approaches for state estimation against sparse sensor attacks are proposed, but they lack optimality guarantees against arbitrary sensor attacks [Mattingley and Boyd, 2010, Farahmand et al., 2011]. Furthermore, detecting malicious components via fault detection and isolation-based methods has been extensively studied for dynamical systems in Pasqualetti et al. (2010) and Fawzi et al. (2014b). However, pinpointing the exact set of malicious components is, in general, a computationally hard problem, as it involves either generating a residue filter for every possible set of malicious sensor [Pasqualetti et al., 2010] or solving an L_0 minimization problem [Fawzi et al., 2014b], both of which are combinatorial in nature.

On another research front line, the design of state estimators, which can tolerate a small portion of the sensory data being altered, has also been extensively studied in Kassam et al. (1985), Maronna et al. (2006), Huber and Ronchetti (2009), and the references therein for more details. Furthermore, in order to measure the security of a robust estimator, Mo and Sinopoli (2015b) propose an estimator that has a minimum mean squared error against the worst-case attacks. However, it must be remarked that the problem of designing a secure state estimator for a dynamical system is much more challenging. Admittedly, the bias injected by an adversary can accumulate in the state estimation, and the adversary can potentially exploit this fact to introduce a large or even unbounded estimation error [Mo and Sinopoli, 2010, 2015a].

Focusing on the problem of bias accumulation in dynamic state estimation, Fawzi et al. (2014b) propose a moving horizon approach for error estimation and correction. In other words, the estimator will only use the measurements from time $k - T + 1$ to time k to estimate the current state $x(k)$, which effectively reduces the dynamic state estimation problem into a static estimation problem. Pajic et al. (2014a, 2015) further develop this approach to systems subject to random or bounded noise. As the main merit of this approach, the static estimation problem can be solved efficiently using ℓ_1 relaxation by exploiting the sparseness of the bias injected by the adversary. However, the sensory data before time $k - T$ is discarded in the moving horizon approach, which may result in a degradation of the estimation performance. To overcome the problem of data discard in the moving horizon approach, Mo and Garone (2016) construct a local state estimator for each sensor, and the historical sensory data can be stored in the local state estimate.

In contrast to prior work based on observable sensors for a linear time-invariant Gaussian system [Mo and Garone, 2016], in this chapter we consider a more general case of the system, which may be unobservable by some sensors. Then we investigate the problem of designing a secure state estimator for the linear time-invariant Gaussian system in the presence of sparse integrity attacks. The set of the malicious sensors is assumed to be fixed over time. The significance of this work is to provide a stable estimator against the (p, m)-sparse attack via local

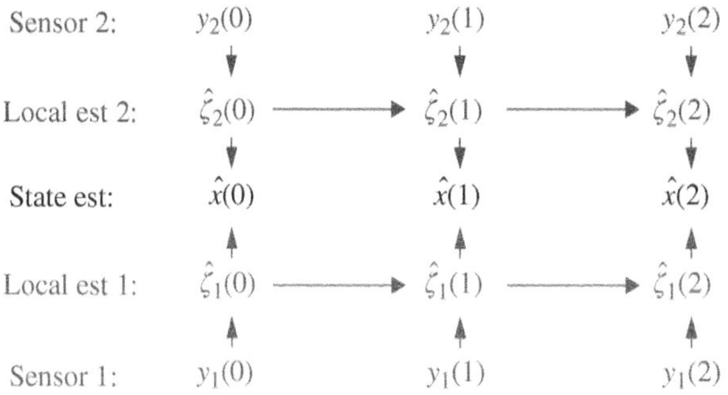

Figure 8.1 The information flow of the proposed filter.

estimators. The structure of our estimate is illustrated in Figure 8.1, and it can be described as follows.

1. For each sensor i, we construct a local state estimator that leverages all the historical measurements from itself to derive a local state estimate.
2. The current global state estimate can then be computed based solely on the current local state estimates using a secure fusion scheme.

The rest of the chapter is organized as follows. In Section 8.2, we formulate the secure state estimation problem. In Section 9.4.1, we prove that the Kalman estimator can be decomposed as a linear combination of local estimators and provide a least-square interpretation for the decomposition. In Section 8.4, a convex optimization-based approach is proposed to derive a more secure state estimate from the local estimates. The performance of the proposed estimator is illustrated via a numerical example in Section 8.5. Finally, Section 8.6 concludes the chapter.

8.2 Problem Formulation

In this section, we introduce the secure state estimation problem. Consider the linear time-invariant system

$$x(k+1) = Ax(k) + w(k), \tag{8.1}$$

where $x(k) \in \mathbb{R}^n$ is the state, and $w(k) \sim \mathcal{N}(0, Q)$ is independent and identically distributed (i.i.d.) Gaussian process noise with zero mean and covariance matrix $Q > 0$. The initial state $x(0) \sim \mathcal{N}(0, \Sigma)$ is assumed to be zero mean Gaussian and is independent from the noise process $\{w(k)\}$.

It is assumed that m sensors are measuring the system and that the measurement from the ith sensor is

$$y_i(k) = C_i x(k) + v_i(k) + a_i(k) = z_i(k) + a_i(k), \tag{8.2}$$

where $y_i(k) \in \mathbb{R}$, $C_i \in \mathbb{R}^{1\times n}$, and $v_i(k) \in \mathbb{R}$ are Gaussian measurement noise. The scalar $a_i(k)$ denotes the bias injected by an adversary. $z_i(k) = C_i x(k) + v_i(k)$ can be regarded as the true measurement without the bias injected by the adversary. Clearly, for a benign sensor i, $a_i(k) = 0$ for all k, while for a compromised sensor i, $a_i(k)$ can be arbitrary. We further assume that the set of compromised sensors remains constant over time.

By defining the aggregated vectors

$$y(k) = \begin{bmatrix} y_1(k) \\ \vdots \\ y_m(k) \end{bmatrix}, \; z(k) = \begin{bmatrix} z_1(k) \\ \vdots \\ z_m(k) \end{bmatrix}, \; C = \begin{bmatrix} C_1 \\ \vdots \\ C_m \end{bmatrix},$$

$$a(k) = \begin{bmatrix} a_1(k) \\ \vdots \\ a_m(k) \end{bmatrix}, \; v(k) = \begin{bmatrix} v_1(k) \\ \vdots \\ v_m(k) \end{bmatrix}, \tag{8.3}$$

we can rewrite (8.2) as

$$y(k) = Cx(k) + v(k) + a(k) = z(k) + a(k). \tag{8.4}$$

We assume that $v(k) \sim \mathcal{N}(0, R)$ with $R > 0$ is i.i.d and independent of the noise process $\{w(k)\}$ and the initial condition $x(0)$. Without loss of generality, we assume (A, C) to be observable.[1]

If all sensors are benign, i.e. $a(k) = 0$ for all k, the optimal state estimator is the classical Kalman filter:

$$\begin{aligned} \hat{x}(k) &= \hat{x}(k|k-1) + K(k)\left[y(k) - C\hat{x}(k|k-1)\right], \\ P(k) &= P(k|k-1) - K(k)CP(k|k-1), \end{aligned}$$

where

$$\hat{x}(k+1|k) = A\hat{x}(k), \; P(k+1|k) = AP(k)A^T + Q,$$
$$K(k) = P(k|k-1)C^T(CP(k|k-1)C^T + R)^{-1},$$

with initial condition

$$\hat{x}(0|-1) = 0, \; P(0|-1) = \Sigma.$$

Since the system is observable, it is well known that the estimation error covariance matrices $P(k)$ and the gain $K(k)$ will converge to

$$P = \lim_{k\to\infty} P(k), \; P_+ = APA^T + Q \tag{8.5}$$

$$K = P_+C^T(CP_+C^T + R)^{-1}. \tag{8.6}$$

---

1 In the case where (A, C) is not observable, we can always perform a Kalman decomposition and only consider the observable space.

Since typically the control system will be running for an extended period of time, we can assume that the Kalman filter is in a steady state or, equivalently, that $\Sigma = P$, and thus the Kalman filter reduces to the following fixed-gain linear estimator:

$$\hat{x}(k+1) = (A - KCA)\hat{x}(k) + Ky(k+1). \tag{8.7}$$

For reasons that will be clearer soon, we will denote with K_i the ith column vector of the matrix $K = [K_1, \dots, K_m]$. Accordingly, (8.7) can be rewritten as

$$\hat{x}(k+1) = (A - KCA)\hat{x}(k) + \sum_{i=1}^{m} K_i y_i(k+1). \tag{8.8}$$

We denote the index set of all sensors as $\mathcal{S} = \{1, 2, \cdots, m\}$. For any index set $\mathcal{I} \subseteq \mathcal{S}$, define the complement set to be $\mathcal{I}^c = \mathcal{S} \backslash \mathcal{I}$. For the attack model discussed in this chapter, we assume that the attacker can only compromise at most p sensors but can arbitrarily choose a_i. Furthermore, we define the collection of all possible index sets of malicious sensors as $\mathcal{C} = \{\mathcal{I} : \mathcal{I} \subseteq \mathcal{S}, |\mathcal{I}| = p\}$. The set of all possible (p, m)-sparse attacks is denoted as $\mathcal{A} = \bigcup_{\mathcal{I} \in \mathcal{C}} \{a(k) : \ \|a_i(k)\| = 0, \ i \in \mathcal{I}^c\}$ for all k.

The goal of this chapter is to propose an algorithm that can estimate the state in such a way that:

1. If no sensor is compromised, i.e. $a_i(k) = 0$ for all i and all k, the estimate coincides with a certain probability to the Kalman estimate obtained in (8.7);
2. The algorithm gives a condition under which the estimator is stable, i.e. with bounded estimation error, against the (p, m)-sparse attack.

To achieve this goal, two results are presented. In the next section, it is shown that, under the mild hypothesis, the estimate of the Kalman filter can be written as a quadratic programming problem involving the estimates generated by a set of local estimators. Then a secure fusion scheme is proposed to replace the quadratic programming.

8.3 Decomposition of the Kalman Filter By Using a Local Estimate

In this section, we introduce a method to decompose the Kalman estimate (8.7) into a linear combination of local state estimates. We will make the following assumption throughout this chapter.

Assumption The matrix A is invertible; $A - KCA$ has n distinct eigenvalues. Moreover, $A - KCA$ and A do not share any eigenvalues. □

Remark 8.1 We assume that (A, C) is observable; then the invertibility of A implies that (A, CA) is also observable. Hence, we can freely assign the poles of $A - KCA$ by choosing a proper gain K. As a result, we can infer that $A - KCA$ can satisfy the condition in *Assumption* 8.1. Otherwise, we can perturb the gain matrix K to enforce the condition, which will only result in a small estimation performance loss if the perturbation is small.

Since $A - KCA$ has distinct eigenvalues, it can be diagonalized as

$$A - KCA = V\Lambda V^{-1}. \tag{8.9}$$

As a result, we can rewrite (8.8) as

$$\left[V^{-1}\hat{x}(k+1)\right] = \Lambda\left[V^{-1}\hat{x}(k)\right] + \sum_{i=1}^{m} V^{-1}K_i y_i(k+1). \tag{8.10}$$

In this chapter, we consider that the system is not necessarily fully observable by the ith sensor, i.e. (A, C_i) may not be observable. In such a case, our goal is to generate m local state estimates $\hat{\zeta}_i(k)$, $i = 1, \dots, m$, such that:

1. Each local estimator generates a stable estimate on the subspace that is observable to the sensor;
2. The Kalman estimate $\hat{x}(k)$ can be recovered as a linear combination of $\hat{\zeta}_i(k)$, i.e.

$$\hat{x}(k) = F_1\hat{\zeta}_1(k) + \dots + F_m\hat{\zeta}_m(k).$$

Now consider the following recursive equation

$$\hat{\zeta}_i(k+1) = \Lambda\hat{\zeta}_i(k) + \mathbf{1}_n y_i(k+1), \tag{8.11}$$

where $\mathbf{1}_n \in \mathbb{R}^{n\times 1}$ is an all-one vector and Λ is defined in (8.9).

Let us choose F_i as

$$F_i = V\text{diag}(V^{-1}K_i),$$

where V is defined in (8.9) and $\text{diag}(V^{-1}K_i)$ is an $n \times n$ diagonal matrix with the jth diagonal entry equal to the jth entry of the vector $V^{-1}K_i$. Comparing (8.11) and (8.10), we can prove that

$$\hat{x}(k) = \sum_{i=1}^{m} F_i\hat{\zeta}_i(k). \tag{8.12}$$

Remark 8.2 It is worth noticing that we do not necessarily need to implement the local estimator on the sensor side since we can let our centralized estimator implement (8.11) for each sensor and then combine them via (8.12). However, depending on the application, it may be advantageous to implement the local estimator on the sensor side to reduce the computational burden of the central estimator.

Next, we will show the relationship between $\hat{\xi}(k)$ and $x(k)$. To this end, let us define the matrices $G_i, i = 1, 2, \cdots, m$, which can be written as

$$G_i = \begin{bmatrix} C_i A(A - \lambda_1 I)^{-1} \\ \vdots \\ C_i A(A - \lambda_n I)^{-1} \end{bmatrix} \in \mathbb{R}^{n \times n},$$

where λ_i is the ith eigenvalues of $A - KCA$ (and Λ). Notice that the inverse of $A - \lambda_i I$ is well defined since A does not share eigenvalues with Λ. The following theorem establishes the connection between $\hat{\xi}_i(k)$ and the state $x(k)$.

Theorem 8.1 *Let $\epsilon_i(k) = G_i x(k) - \hat{\xi}_i(k)$; then*

$$\begin{aligned} \epsilon_i(k+1) &= \Lambda\epsilon_i(k) + (G_i - \mathbf{1}_n C_i)w(k) \\ &\quad -\mathbf{1}_n v_i(k+1) - \mathbf{1}_n a_i(k+1). \end{aligned} \tag{8.13}$$

In other words, $\hat{\xi}_i(k)$ is a stable estimate of $G_i x(k)$ since $A - KCA$ is stable.

Proof: By the definition of $\epsilon_i(k)$, we have

$$\begin{aligned} \epsilon_i(k+1) &= G_i x(k+1) - \hat{\xi}_i(k+1) \\ &= (G_i A - \mathbf{1}_n C_i A)x(k) - \Lambda\hat{\xi}_i(k) \\ &\quad + (G_i - \mathbf{1}_n C_i)w(k) - \mathbf{1}_n v_i(k+1) - \mathbf{1}_n a_i(k+1). \end{aligned}$$

By the definition of G_i, it can be found that

$$G_i A - \mathbf{1}_n C_i A = \begin{bmatrix} C_i A(A - \lambda_1 I)^{-1} A - C_i A \\ \vdots \\ C_i A(A - \lambda_n I)^{-1} A - C_i A \end{bmatrix} = \Lambda G_i. \tag{8.14}$$

Therefore,

$$\begin{aligned} \epsilon_i(k+1) &= \Lambda G_i x(k) - \Lambda\hat{\xi}_i(k) + (G_i - \mathbf{1}_n C_i)w(k) \\ &\quad -\mathbf{1}_n v_i(k+1) - \mathbf{1}_n a_i(k+1) \\ &= \Lambda\epsilon_i(k) + (G_i - \mathbf{1}_n C_i)w(k) \\ &\quad -\mathbf{1}_n v_i(k+1) - \mathbf{1}_n a_i(k+1), \end{aligned}$$

which concludes the proof. □

Remark 8.3 It is worth noticing that if (A, C_i) is not fully observable, then G_i will not be full rank. In fact, by the Cayley-Hamilton theorem, $(A - \lambda_i I)^{-1}$ can be written as a linear combination of $I, A, \ldots, A^{n-1}$. As a result, each row vector of G_i will belong to the observable space of (A, C_i).

The following lemma characterizes a very interesting property of F_i and G_i, for $i = 1, 2, \cdots, m$.

Lemma 8.1 *Suppose that A and $A - KCA$ do not share eigenvalues; then matrices F_i and matrices G_i satisfy the following equation:*

$$\sum_{i=1}^{m} F_i G_i = I. \tag{8.15}$$

Proof: Let us write V^{-1} as the following form

$$V^{-1} = \begin{bmatrix} \alpha_1 & \cdots & \alpha_n \end{bmatrix}^T_{n \times n},$$

where $\alpha_j^T \in \mathbb{R}^{1 \times n}, j = 1, 2 \cdots, n$ denotes the jth row vector of V^{-1}. Then we can compute that

$$F_i G_i = V \text{diag}(V^{-1} K_i) G_i$$
$$= V \begin{bmatrix} \alpha_1^T K_i & & \\ & \ddots & \\ & & \alpha_n^T K_i \\ & & \end{bmatrix} \begin{bmatrix} C_i A(A - \lambda_1 I)^{-1} \\ \vdots \\ C_i A(A - \lambda_n I)^{-1} \\ \end{bmatrix},$$

Since $\alpha_1^T K_i, \cdots, \alpha_n^T K_i$ are scalars, we obtain that

$$\sum_{i=1}^{m} F_i G_i = \sum_{i=1}^{m} V \begin{bmatrix} \alpha_1^T K_i C_i A(A - \lambda_1 I)^{-1} \\ \vdots \\ \alpha_n^T K_i C_i A(A - \lambda_n I)^{-1} \\ \end{bmatrix}_{n \times n}$$
$$= V \begin{bmatrix} \alpha_1^T KCA(A - \lambda_1 I)^{-1} \\ \vdots \\ \alpha_n^T KCA(A - \lambda_n I)^{-1} \\ \end{bmatrix}_{n \times n}.$$

Notice (8.9), and we know that α_j^T is actually a left eigenvector of matrix $A - KCA$ corresponding to eigenvalue λ_j, so we have $\alpha_j^T(A - KCA) = \lambda_j \alpha_j^T$, i.e. $\alpha_j^T KCA = \alpha_j^T(A - \lambda_j I)$ for $j = 1, 2, \cdots, n$. Therefore, $\sum_{i=1}^{m} F_i G_i = VV^{-1} = I$. This completes the proof. □

A Least-Square Interpretation In this subsection, we show that the linear fusion scheme can be interpreted as a least-square problem, which will be used later to derive a secure fusion scheme.

According to the previous discussion, we obtain the state estimation error of the ith local estimation as $\epsilon_i(k)$, i.e.

$$\epsilon_i(k) = G_i x(k) - \hat{\zeta}_i(k),$$

which satisfies the recursive equation (8.13).

Based on the recursive equation (8.13), we let $\epsilon_i(k) = \mu_i(k) + \nu_i(k)$ and define $\mu_i(k)$, $\nu_i(k)$ as follows

$$\begin{aligned} \mu_i(k+1) &= A\mu_i(k) + (G_i - \mathbf{1}_n C_i)w(k) - \mathbf{1}_n v_i(k+1), \\ \nu_i(k+1) &= A\nu_i(k) - \mathbf{1}_n a_i(k+1). \end{aligned} \tag{8.16}$$

where $\mu_i(k)$ can be regarded as the error of the local estimate caused by noise and $\nu_i(k)$ as the error caused by the bias injected by the adversary.

Furthermore, let us define $\tilde{A} \in \mathbb{R}^{mn \times mn}$, $\tilde{\mu}(k) \in \mathbb{R}^{mn}$ as

$$\tilde{A} = \begin{bmatrix} A & & \\ & \ddots & \\ & & A \end{bmatrix}, \quad \tilde{\mu}(k) = \begin{bmatrix} \mu_1(k) \\ \vdots \\ \mu_m(k) \end{bmatrix}. \tag{8.17}$$

Similarly, we can define $\tilde{\epsilon}(k)$ and $\tilde{\nu}(k)$ by stacking $\epsilon_i(k)$ and $\nu_i(k)$ as a big vector.

It can be found that $\tilde{\mu}(k)$ will be Gaussian distributed, and its covariance satisfies the following Lyapunov equation

$$\mathrm{Cov}[\tilde{\mu}(k+1)] = \tilde{A}\mathrm{Cov}[\tilde{\mu}(k)]\tilde{A}^T + \tilde{Q}. \tag{8.18}$$

where

$$\tilde{Q} = \mathrm{Cov}\left[\begin{pmatrix} G_1 - \mathbf{1}_n C_1 \\ \vdots \\ G_m - \mathbf{1}_n C_m \end{pmatrix} w(k)\right] + \mathrm{Cov}\left[\begin{pmatrix} \mathbf{1}_n v_1(k+1) \\ \vdots \\ \mathbf{1}_n v_m(k+1) \end{pmatrix}\right], \tag{8.19}$$

and the covariances are described as

$$\mathrm{Cov}\left[\begin{pmatrix} G_1 - \mathbf{1}_n C_1 \\ \vdots \\ G_m - \mathbf{1}_n C_m \end{pmatrix} w(k)\right] = \begin{bmatrix} G_1 - \mathbf{1}_n C_1 \\ \vdots \\ G_m - \mathbf{1}_n C_m \end{bmatrix} Q \begin{bmatrix} G_1 - \mathbf{1}_n C_1 \\ \vdots \\ G_m - \mathbf{1}_n C_m \end{bmatrix}^T, \tag{8.20}$$

$$\mathrm{Cov}\left[\begin{pmatrix} \mathbf{1}_n v_1(k+1) \\ \vdots \\ \mathbf{1}_n v_m(k+1) \end{pmatrix}\right] = \mathbf{1}_{mn \times mn} \circ \left(R \otimes \mathbf{1}_{n \times n}\right). \tag{8.21}$$

where $\circ$ denotes element-wise matrix multiplication, $\otimes$ is the Kronecker product, and $\mathbf{1}_{mn \times mn}$ is an all-one matrix of size $mn \times mn$.

Next, let us define $\tilde{W}$ as the fixed point [2] of (8.18), i.e.

$$\tilde{W} = \tilde{A}\tilde{W}\tilde{A}^T + \tilde{Q}. \tag{8.22}$$

2 $\tilde{W}$ is well defined since matrix A is strictly stable.

Now, we introduce the following optimization problem:

$$\begin{aligned} \underset{\check{x}(k),\check{\epsilon}(k)}{\text{minimize}} \quad & \frac{1}{2}\check{\epsilon}(k)^T \tilde{W}^{-1} \check{\epsilon}(k) \\ \text{subject to} \quad & \begin{bmatrix} \hat{\zeta}_1(k) \\ \vdots \\ \hat{\zeta}_m(k) \end{bmatrix} = \begin{bmatrix} G_1 \\ \vdots \\ G_m \end{bmatrix} \check{x}(k) - \check{\epsilon}(k). \end{aligned} \tag{8.23}$$

This problem can be interpreted as the problem of finding an estimate $\check{x}(k)$ that minimizes a weighted least square of the error with the local estimates $\hat{\zeta}_i(k)$, where the weighting matrix is related with the covariance of the error of the local estimates.

The following theorem is proposed to establish the connection between the linear fusion scheme (8.12) and the least-square problem (8.23).

Theorem 8.2 *The solution of the least-square problem* (8.23) *is given by*

$$\check{x}(k) = \sum_{i=1}^{m} F_i \hat{\zeta}_i(k) = \hat{x}(k),$$

$$\check{\epsilon}(k) = \left(I - \begin{bmatrix} G_1 \\ \vdots \\ G_m \end{bmatrix} \begin{bmatrix} F_1 & \dots & F_m \end{bmatrix} \right) \tilde{\epsilon}(k).$$

Remark 8.4 It should be mentioned that the results and proofs presented in this section are purely algebraic. Therefore, the result can be easily generalized to other linear fixed-gain estimators and other noise models.

It is worth noticing that the linear fusion scheme (8.12) (or, equivalently, the least-square problem (8.23)) is not secure in the sense that if sensor i is compromised, the adversary can manipulate $\hat{\zeta}_i(k)$ by injecting the bias $a_i(k)$ into the measurements $y_i(k)$. Therefore, the adversary can potentially change the Kalman estimate arbitrarily. In the next section, to address the security challenges, we modify (8.23) by adding an ℓ_1 penalty to ensure the stability of the state estimation in the presence of malicious sensors.

8.4 A Secure Information Fusion Scheme

In this section, we consider two scenarios of the attack model and propose a convex optimization-based approach to combine the local estimate into a more secure state estimate.

Notice that the error $\epsilon_i(k)$ can be decomposed as the error caused by the noise $\mu_i(k)$ and the error caused by the bias injected by the adversary $v_i(k)$. As a result, we introduce the following secure fusion scheme based on LASSO [Tibshirani, 1996]

$$\underset{\check{x}_s(k),\check{\mu}(k),\check{v}(k)}{\text{minimize}} \frac{1}{2}\check{\mu}(k)^T\tilde{W}^{-1}\check{\mu}(k) + \gamma\|\check{v}(k)\|_1 \tag{8.24}$$
$$\text{subject to } \hat{\zeta}_i(k) = G_i\check{x}_s(k) - \check{\mu}_i(k) - \check{v}_i(k),\ \forall i \in S,$$

where $\check{x}_s(k)$ is the secure state estimation. γ is a constant chosen by the system operator. $\check{\mu}(k)$, $\check{v}(k)$ are defined as

$$\check{\mu}(k) = \begin{bmatrix} \check{\mu}_1(k) \\ \vdots \\ \check{\mu}_m(k) \end{bmatrix},\ \check{v}(k) = \begin{bmatrix} \check{v}_1(k) \\ \vdots \\ \check{v}_m(k) \end{bmatrix}.$$

The following lemma is proposed to characterize the solution of the optimization problem.

Lemma 8.2 *Let $\check{x}_s(k)$, $\check{\mu}(k)$, $\check{v}(k)$ be the minimizer for the optimization problem* (8.24). *Let $\check{x}(k)$, $\check{\epsilon}(k)$ be the minimizer for the least-square problem* (8.23). *Then the following statements hold:*

1. *The following inequality holds:*

$$\|\tilde{W}^{-1}\check{\mu}(k)\|_\infty \leq \gamma. \tag{8.25}$$

2. *If $\|\tilde{W}^{-1}\check{\epsilon}(k)\|_\infty \leq \gamma$, then*

$$\check{x}_s(k) = \check{x}(k) = \hat{x}(k),\ \check{\mu}(k) = \check{\epsilon}(k),\ \check{v}(k) = 0.$$

Proof: Let us prove (8.25) with the reduction to absurdity. Suppose that

$$\|\tilde{W}^{-1}\check{\mu}(k)\|_\infty > \gamma.$$

Therefore, we can find a vector η, such that

$$\eta^T\tilde{W}^{-1}\check{\mu}(k) > \gamma,$$

with $\|\eta\|_1 = 1$. For any $\alpha > 0$, $\check{x}_s(k)$, $\check{\mu}(k) - \alpha\eta$, and $\check{v}(k) + \alpha\eta$ will also be a feasible solution for (8.24). The corresponding cost function can be calculated as

$$\begin{aligned} &\tfrac{1}{2}(\check{\mu}(k) - \alpha\eta)^T\tilde{W}^{-1}(\check{\mu}(k) - \alpha\eta) + \gamma\|\check{v}(k) + \alpha\eta\|_1 \\ &\quad\leq \tfrac{1}{2}\check{\mu}(k)^T\tilde{W}^{-1}\check{\mu}(k) + \gamma\|\check{v}(k)\|_1 \\ &\quad + \alpha\left(\gamma\|\eta\|_1 - \eta^T\tilde{W}^{-1}\check{\mu}(k)\right) + \tfrac{1}{2}\alpha^2\eta^T\tilde{W}^{-1}\eta. \end{aligned}$$

Notice that

$$\gamma\|\eta\|_1 - \eta^T \tilde{W}^{-1} \check{\mu}(k) < 0.$$

Choose $\alpha < 2[\eta^T \tilde{W}^{-1} \check{\mu}(k) - \gamma](\eta^T \tilde{W}^{-1} \eta)^{-1}$, and we have

$$\frac{1}{2}(\check{\mu}(k) - \alpha\eta)^T \tilde{W}^{-1} (\check{\mu}(k) - \alpha\eta) + \gamma\|\check{v}(k) + \alpha\eta\|_1$$
$$< \frac{1}{2}\check{\mu}(k)^T \tilde{W}^{-1} \check{\mu}(k) + \gamma\|\check{v}(k)\|_1,$$

which contradicts with the optimality of $\check{x}_s(k)$, $\check{\mu}(k)$ and $\check{v}(k)$. Therefore, (8.25) must hold.

Since the second statement can be clearly proved by utilizing KKT conditions, the detailed proof is omitted due to space limitations. □

We now consider two scenarios:

1. All sensors are benign, and the system is operating normally.
2. p sensors are compromised.

The following two theorems characterize the performance of the secure fusion scheme (8.24) for each scenario.

Theorem 8.3 *Suppose that all the sensors are benign, i.e. $a(k) = 0$ for all k. We conclude that $\check{x}_s(k) = \check{x}(k) = \hat{x}(k)$ if the following inequality holds:*

$$\left\| \tilde{W}^{-1} \left(I - \begin{bmatrix} G_1 \\ \vdots \\ G_m \end{bmatrix} \begin{bmatrix} F_1 & \dots & F_m \end{bmatrix} \right) \tilde{\epsilon}(k) \right\|_\infty \le \gamma. \tag{8.26}$$

Proof: It can be seen that this theorem is a direct consequence of *Lemma* 8.2 and *Theorem* 8.2, so the detailed proof is omitted here. □

Remark 8.5 If all sensors are benign, the local estimation error $\epsilon_i(k)$ will be zero-mean Gaussian distributed and

$$\lim_{k\to\infty} \text{Cov}(\tilde{\epsilon}(k)) = \tilde{W}.$$

If the system is operating long enough, we have $\text{Cov}(\tilde{\epsilon}(k)) \approx \tilde{W}$, and we can compute the probability that the inequality (8.26) holds, i.e. the probability that the secure state estimate equals the optimal Kalman estimate. Moreover, increasing γ will increase the likelihood that the secure estimation equals the Kalman estimation during normal operation.

For the second scenario, we consider that p ($p < m$) sensors are compromised. The stability of the proposed secure estimator is characterized by the following theorem, the proof of which is reported in the appendix for the sake of legibility:

Theorem 8.4 *Suppose that p ($p < m$) sensors are compromised, then the secure state estimate $\check{x}_s(k)$ is stable against the (p, m)-sparse attack if the following inequality holds for all $u \neq 0$:*

$$\sum_{i \in \mathcal{I}} \|G_i u\|_1 < \sum_{i \in \mathcal{I}^c} \|G_i u\|_1, \ \forall \mathcal{I} \in \mathcal{C}.$$

8.5 Numerical Example

In this section, we demonstrate our proposed secure estimation via a numerical example. We assume the following parameters for our system:

$$A = \begin{bmatrix} 1 & 0 \\ 0 & -1 \end{bmatrix}, \ C = \begin{bmatrix} C_1 \\ C_2 \\ C_3 \\ C_4 \\ C_5 \end{bmatrix} = \begin{bmatrix} 1 & 0 \\ 0 & 1 \\ 1 & 1 \\ 1 & -2 \\ -1 & 1 \end{bmatrix}, \ Q = I, \ R = I.$$

We can verify that the system is not fully observable by the first sensor and the second sensor. In this situation, we will compute the secure estimator and hope to achieve good performance for two scenarios.

The optimal steady-state Kalman gain K and estimation covariance P matrices are given by

$$K = \begin{bmatrix} 0.2311 & 0.0575 & 0.2886 & 0.1162 & -0.1736 \\ 0.0575 & 0.1412 & 0.1987 & -0.2250 & 0.0838 \end{bmatrix},$$

$$P = \begin{bmatrix} 0.2311 & 0.0575 \\ 0.0575 & 0.1412 \end{bmatrix}.$$

The corresponding $A - KCA$ matrix has eigenvalues at 0.1802 and −0.1161. As a result, we can derive the matrices G_i as follows:

$$G_1 = \begin{bmatrix} 0.8960 & 0 \\ 1.2199 & 0 \end{bmatrix}, G_2 = \begin{bmatrix} 0 & 1.1314 \\ 0 & 0.8473 \end{bmatrix}, G_3 = \begin{bmatrix} 0.8960 & 1.1314 \\ 1.2199 & 0.8473 \end{bmatrix},$$

$$G_4 = \begin{bmatrix} 0.8960 & -2.2627 \\ 1.2199 & -1.6946 \end{bmatrix}, G_5 = \begin{bmatrix} -0.8960 & 1.1314 \\ -1.2199 & 0.8473 \end{bmatrix}.$$

Then we consider the following two scenarios:

1. All sensors are benign;
2. The first sensor is under attack, and $a_1(k) = 100$ for all k.

We compute the empirical mean squared error (MSE) of the secure estimator for each scenario and different choices of γ. Notice that when all the sensors are benign, the optimal Kalman estimator has an MSE which can be computed as $\mathrm{tr}(P) = 0.3723$. As a result, we define the normalized MSE as the MSE divided by 0.3723. Figure 8.2 illustrates the normalized MSE of the proposed secure

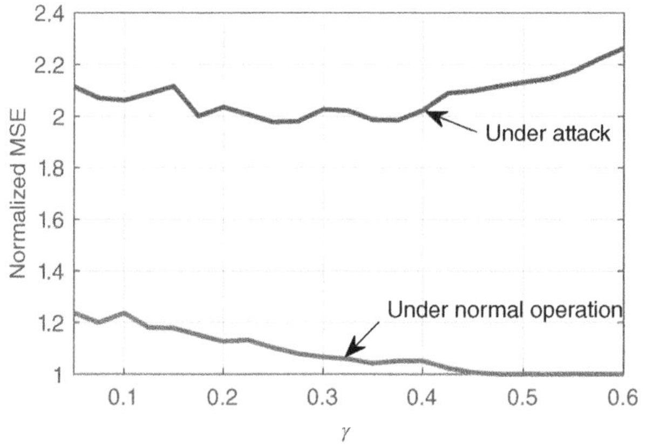

Figure 8.2 The normalized mean squared error of the secure estimator versus different choices of γ.

estimator versus γ. The light gray line indicates the normalized MSE when all sensors are benign, while the dark gray line indicates the normalized MSE when sensor 1 is malicious. It can be seen that when $\gamma \geq 0.45$, the secure estimator achieves roughly the same estimation performance as the optimal Kalman estimator under normal operation. On the other hand, if sensor 1 is malicious, the MSE achieves the minimum at around $\gamma = 0.25$.

8.6 Conclusion

In this chapter, we have studied the problem of estimating the state of a linear time-invariant Gaussian system in the presence of sparse integrity attacks. The attacker can control p out of m sensors and arbitrarily change the measurements. Moreover, the system may be unobservable by some sensors. Under mild assumptions, we have proven that we can decompose the optimal Kalman estimate as a weighted sum of local state estimates. In order to find a stable state estimator against the (p, m)-sparse attack, we have introduced a convex optimization-based approach to combine the local estimate into a more secure state estimate. Finally, a numerical example illustrated that our secure estimator achieves good performance under both normal operation and attack scenarios.

8.7 Appendix: Proof of Theorem 8.2

Before proving the theorem, we will prove the following lemma, which will be used in the proof of *Theorem* 8.2.

Lemma 8.3 *Let K be the steady-state Kalman gain defined in (8.6). For $j = 1, 2, \cdots, m$, the following Lyapunov equation holds*

$$PG_j^T = (A - KCA)PG_j^T \Lambda + (I - KC)Q(G_j - \mathbf{1}_n C_j)^T + \sum_{i=1}^{m} r_{ij} K_i \mathbf{1}_n^T, \tag{8.27}$$

where P is defined in (8.5).

Proof: Based on the result of Lemma 4 in Mo and Garone (2016), for any $L_{j,o}$ we know that

$$P = (A - KCA)P(A - L_{j,o}C_jA)^T + (I - KC)Q(I - L_{j,o}C_j)^T + \sum_{i=1}^{m} r_{ij} K_i L_{j,o}^T.$$

Here we choose $L_{j,o}$ such that $L_{j,o}^T G_{j,o}^T = 1$, where $o = 1, 2, \cdots, n$ and $G_{j,o} = C_jA(A - \lambda_o I)^{-1}$ represent the oth row of matrix G_j. Since $G_{j,o}$ is a non-zero row vector of matrix G_j, we can always find a column vector $L_{j,o}$ to satisfy the condition $G_{j,o}L_{j,o} = 1$. Hence we can obtain that

$$PG_{j,o}^T = (A - KCA)P(A^T G_{j,o}^T - A^T C_j^T) + (I - KC)Q(G_{j,o}^T - C_j^T) + \sum_{i=1}^{m} r_{ij} K_i.$$

Due to $G_j^T = [G_{j,1}^T \ \cdots \ G_{j,o}^T \ \cdots \ G_{j,n}^T]$, it implies that

$$PG_j^T = (A - KCA)P(A^T G_j^T - A^T C_j^T \mathbf{1}_n^T) + (I - KC)Q(G_j - \mathbf{1}_n C_j)^T + \sum_{i=1}^{m} r_{ij} K_i \mathbf{1}_n^T.$$

According to (8.14), we can conclude that the matrix equation (8.27) holds for $j = 1, 2, \cdots, m$. □

Proof: (*Proof of Theorem 8.2*) We rewrite the matrix $\tilde{W}$ in a block diagonal form

$$\tilde{W} = \begin{bmatrix} \tilde{W}_{11} & \cdots & \tilde{W}_{1m} \\ \vdots & \ddots & \vdots \\ \tilde{W}_{m1} & \cdots & \tilde{W}_{mm} \end{bmatrix},$$

where each $\tilde{W}_{ij} \in \mathbb{R}^{n \times n}$. As a result, by (8.22), we know that $\tilde{W}_{ij}$ satisfies

$$\tilde{W}_{ij} = \Lambda \tilde{W}_{ij} \Lambda + (G_i - \mathbf{1}_n C_i)Q(G_j - \mathbf{1}_n C_j)^T + r_{ij} \mathbf{1}_n \mathbf{1}_n^T,$$

where r_{ij} is the element of the matrix R in the ith row and jth column. Since $F_i\mathbf{1}_n = K_i$ and

$$F_i\Lambda = V\text{diag}(V^{-1}K_i)\Lambda$$
$$= V\Lambda V^{-1}V\text{diag}(V^{-1}K_i) = (A - KCA)F_i,$$

it follows that

$$F_i\tilde{W}_{ij} = (A - KCA)F_i\tilde{W}_{ij}\Lambda$$
$$+ (F_iG_i - K_iC_i)Q(G_j - \mathbf{1}_nC_j)^T + r_{ij}K_i\mathbf{1}_n^T.$$

Therefore, let $\tilde{S}_j = \sum_{i=1}^m F_i\tilde{W}_{i,j}$, by *Lemma* 8.1 and $\sum_{i=1}^m(F_iG_i - K_iC_i) = I - KC$, we can conclude that $\tilde{S}_j$ satisfies the following recursive equation:

$$\tilde{S}_j = (A - KCA)\tilde{S}_j\Lambda + (I - KC)Q(G_j - \mathbf{1}_nC_j)^T$$
$$+ \sum_{i=1}^m r_{ij}K_i\mathbf{1}_n^T. \tag{8.28}$$

Hence, by *Lemma* 8.3, $\tilde{S}_j = PG_j^T$ for all $j = 1, \dots, m$, which implies that

$$\begin{bmatrix} F_1 & \cdots & F_m \end{bmatrix} \tilde{W} = P \begin{bmatrix} G_1^T & \cdots & G_m^T \end{bmatrix}. \tag{8.29}$$

On the other hand, according to the result of least-square estimation, it is easy to show that the optimal solution of (8.23) is given by

$$\check{x}(k) = (\mathbf{G}^T\tilde{W}^{-1}\mathbf{G})^{-1}\mathbf{G}^T\tilde{W}^{-1}\begin{bmatrix} \hat{\zeta}_1(k) \\ \vdots \\ \hat{\zeta}_m(k) \end{bmatrix},$$

where $\mathbf{G}^T = \begin{bmatrix} G_1^T & \cdots & G_m^T \end{bmatrix}$. By (8.29), we have that

$$\mathbf{G}^T\tilde{W}^{-1}\mathbf{G} = P^{-1}\begin{bmatrix} F_1 & \cdots & F_m \end{bmatrix}\mathbf{G} = P^{-1}.$$

Therefore, it can be obtained that

$$\check{x}(k) = \begin{bmatrix} F_1 & \cdots & F_m \end{bmatrix}\begin{bmatrix} \hat{\zeta}_1(k) \\ \vdots \\ \hat{\zeta}_m(k) \end{bmatrix} = \hat{x}(k).$$

According to the definition of $\tilde{e}(k)$, we can get that

$$\begin{bmatrix} \hat{\zeta}_1(k) \\ \vdots \\ \hat{\zeta}_m(k) \end{bmatrix} = \begin{bmatrix} G_1 \\ \vdots \\ G_m \end{bmatrix} x(k) - \tilde{e}(k). \tag{8.30}$$

On the other hand, from the optimization problem (8.23), we know that

$$\begin{bmatrix} \hat{\zeta}_1(k) \\ \vdots \\ \hat{\zeta}_m(k) \end{bmatrix} = \begin{bmatrix} G_1 \\ \vdots \\ G_m \end{bmatrix}\begin{bmatrix} F_1 & \cdots & F_m \end{bmatrix}\begin{bmatrix} \hat{\zeta}_1(k) \\ \vdots \\ \hat{\zeta}_m(k) \end{bmatrix} - \check{e}(k). \tag{8.31}$$

From (8.30) and (8.31), it follows that

$$\breve{\epsilon}(k) = \left(\begin{bmatrix} G_1 \\ \vdots \\ G_m \end{bmatrix} [F_1 \dots F_m] - I \right) \begin{bmatrix} \hat{\zeta}_1(k) \\ \vdots \\ \hat{\zeta}_m(k) \end{bmatrix}$$

$$= \left(I - \begin{bmatrix} G_1 \\ \vdots \\ G_m \end{bmatrix} [F_1 \dots F_m] \right) \tilde{\epsilon}(k).$$

This completes the proof. □

8.8 Proof of Theorem 8.4

Before proving *Theorem* 8.4, we need the following lemma:

Lemma 8.4 *Considering the following optimization problem:*

$$e = \arg\min \sum_{i=1}^{m} \|\eta_i - G_i e\|_1 = g(\eta_1, \dots, \eta_m). \tag{8.32}$$

If the following inequality holds for all $u \neq 0$,

$$\sum_{i \in \mathcal{I}} \|G_i u\|_1 < \sum_{i \in \mathcal{I}^c} \|G_i u\|_1, \ \forall \mathcal{I} \in \mathcal{C},$$

then the following statements hold:

1. *For $i = 1, 2, \cdots, m$,*

$$\|e\|_1 \leq \frac{2m \max_i \|\eta_i\|_1}{\delta_1}.$$

2. *Assuming that*

$$e' = g(\eta'_1, \dots, \eta'_m),$$

where $\eta_i - \eta'_i = v_i$ and at most p v_is are non-zero, then $\|e - e'\|_1$ is bounded,[3] *i.e.*

$$\|e - e'\|_1 \leq \frac{2 \max_i \|\eta'_i\|_1 + \delta_2}{\delta_2} + \frac{2m \max_i \|\eta'_i\|_1}{\delta_1},$$

where

$$\delta_1 = \min_{\|u\|_1 = 1} \sum_{i=1}^{m} \|G_i u\|_1,$$

$$\delta_2 = \frac{1}{m} \min_{\|u\|_1 = 1} \min_{\mathcal{I} \in \mathcal{C}} \left(\sum_{i \in \mathcal{I}^c} \|G_i u\|_1 - \sum_{i \in \mathcal{I}} \|G_i u\|_1 \right).$$

3 Notice that the $\|e - e'\|$ is bounded regardless of $v_1, \dots, v_m$.

Proof:

1. Since e is the optimal solution, we have

$$\sum_{i=1}^{m} \|\eta_i - G_i \times 0\|_1 \geq \sum_{i=1}^{m} \|\eta_i - G_i e\|_1$$
$$\geq \sum_{i=1}^{m} \|G_i e\|_1 - \sum_{i=1}^{m} \|\eta_i\|_1.$$

Then we know $\sum_{i=1}^{m} \|G_i e\|_1 \leq 2 \sum_{i=1}^{m} \|\eta_i\|_1$. It implies that

$$\left(\min_{\|e\|_1=1} \sum_{i=1}^{m} \|G_i e\|_1 \right) \|e\|_1 \leq 2 \sum_{i=1}^{m} \|\eta_i\|_1 \leq 2m \max_i \|\eta_i\|_1.$$

Since $\sum_{i\in\mathcal{I}} \|G_i u\|_1 < \sum_{i\in\mathcal{I}^c} \|G_i u\|_1$ for all $u \neq 0$, we know $\sum_{i=1}^{m} \|G_i u\|_1 > 0$, and we can conclude the proof.

2. Let us define the function $f_i(\vartheta) = \|\vartheta\|_1$. Then it is easily found that the optimization problem (8.32) can be written as

$$e = \arg\min_{e} \sum_{i=1}^{m} f_i(\eta_i - G_i e) = g(\eta_1, \ldots, \eta_m).$$

So we have that

$$e' = \arg\min_{e'} \sum_{i=1}^{m} f_i(\eta_i' - G_i e') = g(\eta_1', \ldots, \eta_m').$$

Now we can choose constant

$$N_i(\eta_i', \delta) = \frac{2\|\eta_i'\|_1}{\delta_2}$$

such that for all $t \geq N_i(\eta_i', \delta_2)$, the following inequality holds:

$$\frac{1}{t}[f_i(tG_i u - \eta_i') - f_i(-\eta_i')]$$
$$= \frac{1}{t}(\|tG_i u - \eta_i'\|_1 - \|\eta_i'\|_1) \geq \|G_i u\|_1 - \delta_2.$$

Based on the result of Han et al. (2015), we can obtain that

$$f_i((t+1)G_i u - \eta_i') - f_i(tG_i u - \eta_i') \geq \|G_i u\|_1 - \delta_2$$

for all $t \geq \max_{1\leq i\leq m} N_i(\eta_i', \delta_2)$ and $\|u\|_1 = 1$.

For benign sensors, $v_i = 0$, i.e. $\eta_i' = \eta_i$, we have that

$$\sum_{i\in\mathcal{I}^c} f_i(\eta_i - (t+1)G_i u) - \sum_{i\in\mathcal{I}^c} f_i(\eta_i - tG_i u)$$
$$\geq \sum_{i\in\mathcal{I}^c} \|G_i u\|_1 - (m-p)\delta_2. \tag{8.33}$$

For malicious sensors, we obtain that

$$\sum_{i\in\mathcal{I}} f_i(\eta_i - tG_iu) - \sum_{i\in\mathcal{I}} f_i(\eta_i - (t+1)G_iu) \leq \sum_{i\in\mathcal{I}} \|G_iu\|_1. \tag{8.34}$$

Then from (8.33) and (8.34), we know that

$$\sum_{i\in\mathcal{S}} f_i(\eta_i - (t+1)G_iu) - \sum_{i\in\mathcal{S}} f_i(\eta_i - tG_iu)$$
$$\geq \sum_{i\in\mathcal{I}^c} \|G_iu\|_1 - \sum_{i\in\mathcal{I}} \|G_iu\|_1 - (m-p)\delta_2 > 0.$$

It implies that any point

$$\|e\|_1 \geq \frac{\max_{1\leq i\leq m} 2\|\eta_i'\|_1}{\delta_2} + 1$$

cannot be the solution of the optimization problem (8.32) since there exists a better point $\frac{\|e\|_1-1}{\|e\|_1}e$. In other words, for the optimal solution, we must have

$$\|e\|_1 = \|g(\eta_1, \cdots, \eta_m)\|_1 \leq \frac{\max_{1\leq i\leq m} 2\|\eta_i'\|_1}{\delta_2} + 1.$$

Therefore, it follows that

$$\|e - e'\|_1 \leq \|e\|_1 + \|e'\|_1$$
$$\leq \frac{2\max_i \|\eta_i'\|_1 + \delta_2}{\delta_2} + \frac{2m\max_i \|\eta_i'\|_1}{\delta_1}.$$

□

We are now ready to prove *Theorem* 8.4.

Proof: Let us define $e_s(k) = x(k) - \check{x}_s(k)$. By subtracting G_ix on both sides of the constraint equation in optimization problem (8.24), it is easy to prove that $e_s(k)$ is the solution to the following minimization problem:

$$\underset{e_s(k),\check{\mu}(k),\check{v}(k)}{\text{minimize}} \frac{1}{2}\check{\mu}(k)^T\tilde{W}^{-1}\check{\mu}(k) + \gamma\|\check{v}(k)\|_1 \tag{8.35}$$
$$\text{subject to } \mu_i(k) + v_i(k) = G_ie_s(k) + \check{\mu}_i(k) + \check{v}_i(k), \forall i \in \mathcal{S}.$$

Let us define

$$\eta_i(k) = \mu_i(k) + v_i(k) - \check{\mu}_i(k), \ \eta_i'(k) = \mu_i(k) - \check{\mu}_i(k). \tag{8.36}$$

According to the optimization problem (8.32), apparently we know that

$$e_s(k) = g(\eta_1, \ldots, \eta_m).$$

Now by (2) of *Lemma* 8.4 for all k, we obtain that

$$\|e_s(k) - g(\eta_1'(k), \dots, \eta_m'(k))\|_1 \leq \frac{2\max_i \|\eta_i'(k)\|_1 + \delta_2}{\delta_2} + \frac{2m\max_i \|\eta_i'(k)\|_1}{\delta_1}.$$

Furthermore, by (1) of *Lemma* 8.4 for all k, we have

$$\|g(\eta_1'(k), \dots, \eta_m'(k))\|_1 \leq \frac{2m\max_i \|\eta_i'(k)\|_1}{\delta_1}.$$

Therefore, it implies that

$$\begin{aligned}
&\|x(k) - \check{x}_s(k)\|_1 = \|e_s(k)\|_1 \\
&\leq \|e_s(k) - g(\eta_1'(k), \dots, \eta_m'(k))\|_1 + \|g(\eta_1'(k), \dots, \eta_m'(k))\|_1 \\
&\leq \frac{2\max_i \|\eta_i'(k)\|_1 + \delta_2}{\delta_2} + \frac{4m\max_i \|\eta_i'(k)\|_1}{\delta_1}.
\end{aligned} \tag{8.37}$$

According to the inequality (8.25), we know that $\check{\mu}_i(k)$ is bounded. So we can claim that the bound of the estimation error is irrelevant to the attack.

Since the estimation error is bounded as (8.37), we can conclude that the secure estimator $\check{x}_s(k)$ is stable against the (p, m)-sparse attack. This completes the proof. □

9

Secure Dynamic State Estimation for AHRS

9.1 Introduction

Attitude information, which includes roll, pitch, and yaw in three-dimensional space, is always required for perception, planning, and control systems on autonomous vehicles (AVs). With the development of microelectromechanical systems, affordable sensors with excellent performance have been developed for attitude estimation. Attitude and heading reference systems (AHRSs) are widely used in, but not limited to, fields such as underwater navigation, artillery collimation, aircraft guidance, satellite communication, and so on [Huang and Chiang, 2008, Wu et al., 2013, Hu and Sun, 2015, Chen et al., 2016]. An AHRS aims to track the attitude of an object by using measurements provided by an inertial measurement unit (IMU) consisting of a triad of accelerometers and gyroscopes. As an integrated sensor, the AHRS outputs not only raw measurements (such as linear acceleration and angular velocity) but also a fused pose. Thus it has become an indispensable sensor on AVs [Wang et al., 2018b].

AHRSs can be designed in various ways based on the physical principles behind the sensor [Suh, 2018]. In environments where a constant vector field is available, attitude is typically obtained by comparing a set of vectors measured in the body-fixed coordinate frame with a set of reference vectors in the global coordinate frame [Markley and Mortari, 2000]. The whole process can be modeled as a state estimation problem, where general filtering-based approaches also apply [Wu et al., 2016, Costanzi et al., 2016, Inoue et al., 2016, Tong et al., 2018]. With the popularization of the Internet of Vehicles (IoV), cybersecurity for AVs has become a noteworthy issue [Parkinson et al., 2017]. Increasingly, vehicles are connected online such that fleet management and monitoring [Gerland, 1993, Parent, 2007], and formation control and coordination [Tanner et al., 2003, Ghabcheloo et al., 2006, Ge et al., 2016] can be applied. Communication standards and technologies have been established and developed to provide support for the growing needs of networked vehicles [Shariatmadari et al., 2015, Gerla et al., 2014,

Multimodal Perception and Secure State Estimation for Robotic Mobility Platforms, First Edition.
Xinghua Liu, Rui Jiang, Badong Chen, and Shuzhi Sam Ge.

Biswas et al., 2006]. AVs may face serious security threats if no secure measures are considered and implemented. Thus, cybersecurity concerns push us to study the secure estimation problem in AHRSs under attack, which remains an ongoing research issue.

Motivated by such discussions, this chapter focuses on securely estimating attitude for low-cost AHRSs. By utilizing a convex optimization-based approach, a secure filter scheme is first established for the attitude estimation and attack models. We introduce a computationally efficient method to decompose the Kalman estimate into a linear combination of estimates generated by a set of local estimators; then a secure fusion scheme is presented to replace the linear fusion scheme. In addition, we provide a sufficient condition under which the proposed estimator is stable against sparse attacks. The highlights of this chapter can be summarized as follows:

1. Based on the low dynamics assumption, a secure attitude estimation framework is presented in which the AHRS measuring model and the attack model are established for AVs.
2. If all the elements of the model state are benign, the proposed secure estimator can coincide with the Kalman estimator with a certain probability. This property makes the proposed approach compatible with existing sensors such as AHRSs in this work.
3. If p elements of the model state are compromised, the proposed secure estimator still gives a stable estimate against a sparse attack. The bounded estimation error is theoretically ensured such that the worst case can be estimated and tested in advance when designing sensor signal processing algorithms.

In the next section, we review the related work on attitude estimation and secure state estimation. Section 9.3 formulates the problem by deriving the estimated attitude from heading reference measurements and breaking down the problem such that assumptions in the proposed secure estimator can be satisfied. The secure estimator design and detailed mathematical proofs are presented in Section 9.4. Experimental validation and results are shown in Section 9.5. Finally, Section 9.6 concludes the chapter.

9.2 Related Work

9.2.1 Attitude Estimation

Given an ambient vector field, the attitude in the global frame can be computed based on the measurement disparity between the global frame and the local frame. This problem was first modeled as Wahba's problem [Wahba, 1965]. To solve Wahba's problem, three-axis attitude determination (TRIAD) [Shuster and Oh, 1981] was proposed such that the attitude can be analytically obtained from two corresponding vector measurements at the cost of discarding partial measurements. To fully utilize measurements, optimization-oriented approaches such as QUaternion ESTimator (QUEST) and its variants [Yun et al., 2008] have been proposed to shorten the execution time. Although this chapter does not stress on optimizing Wahba's problem, Section 9.3 will present the secure estimation process and introduce Wahba's problem and QUEST in detail.

9.2.2 Secure State Estimation

Cyber-physical systems (CPSs) that integrate information technology infrastructures with physical processes are ubiquitous in modern societies and enable the design and implementation of large, complex systems in a remote operation scenario [Hespanha et al., 2007]. Examples of CPSs include smart grids, autonomous automobile systems, process control systems, and robotics systems [Khaitan and McCalley, 2015]. However, the tight coupling between "cyber" components and "physical" processes often leads to systems where increased sophistication comes at the expense of increased vulnerability and security weaknesses. Any successful attacks on CPSs may cause enormous losses, including leakage of classified information, suspension of industrial processes, infrastructure destruction, and even loss of human lives. Illustrative examples are Iran's nuclear centrifuge accident [Farwell and Rohozinski, 2011] and the western Ukraine blackout [BBC, 2016].

From the controller's point of view, in recent years, researchers have investigated how we can securely estimate the state of a dynamic system from a set of noisy and maliciously corrupted sensor measurements [Pasqualetti et al., 2013, Fawzi et al., 2014a, Pajic et al., 2014b, Shoukry et al., 2017b, Ma et al., 2017a]. Pasqualetti et al. (2013) proposed a mathematical framework for CPSs, attacks, and monitors. They studied the fundamental monitoring limitations for linear systems (especially power networks) and characterized the vulnerability of linear systems to cyber-physical attacks using graph theory. To solve the problem of bias accumulation in dynamic state estimation, the combinatorial optimization-based approach is proposed in Fawzi et al. (2014a), Pajic et al. (2014b), and Shoukry et al. (2017b). These approaches usually iterate over all possible attack scenarios, i.e. the set of compromised sensors and actuators, and decide which one is most likely given the sensory data. The main merit is that the stability of the estimator is guaranteed, i.e. an upper bound for the estimation error covariance is guaranteed. However, since the number of possible attack scenarios is combinatorial, these approaches often scale poorly and may not be usable for AVs due to real-time constraints.

9.2.3 Secure Attitude Estimation

As an important application area of CPSs, AVs have become the focus of researchers, and many important results have been reported in the literature, please refer to Lam et al. (2016), Kothari et al. (2017), and Chen et al. (2018) and the references therein. Motivated by the principle of secure state estimation for

CPSs, in this chapter we consider the problem of secure attitude estimation for AHRSs measurement in AV systems. We focus on a linear dynamical system based on the low dynamics assumption and model the attack as a sparse vector added to the measurement vector. We make no assumptions regarding the magnitude, statistical description, or temporal evolution of the attack vector. To the best of the authors' knowledge, the problem of secure attitude estimation for AVs has not been properly investigated previously, nor has a robust attitude estimator been designed against attacks on AVs. Such a situation motivated the present study.

9.3 Attitude Estimation Using Heading References

In this section, we formulate the heading references-based attitude estimation problem and propose a secure state estimation framework that collects raw measurements from heading reference sensors and outputs the estimated attitude.

9.3.1 Attitude Estimation from Vector Observations

Assume we have a sensor that measures ambient vector field in local Cartesian coordinates, as shown in Figure 9.1. Without loss of generality, we suppose that the heading reference is the ambient vector field denoted by $\mathbf{v}$, and the ith measurements of $\mathbf{v}$ in the global frame and the body frame are represented as ${}_{\mathrm{g}}\mathbf{y}_i$ and ${}_{\mathrm{b}}\mathbf{v}_i$, respectively. Since the heading measurement always satisfies the constraint $\|{}_{\mathrm{g}}\mathbf{v}_i\| = \|{}_{\mathrm{b}}\mathbf{v}_i\|$, a single heading measurement provides constraints on two degrees of freedom. As there are three degrees of freedom for body rotation, at least two measurements are required to obtain a unique attitude estimation. By assuming that $\mathbf{v}$ is invariant with regard to time and space, the noise-free attitude estimation problem containing N sensors is to find a rotation matrix ${}^{\mathrm{b}}_{\mathrm{g}}\mathrm{R}$ that satisfies

$$
{}_{\mathrm{g}}\mathbf{v}_i = {}^{\mathrm{b}}_{\mathrm{g}}\mathrm{R}\,{}_{\mathrm{b}}\mathbf{v}_i \;\text{ or }\; {}_{\mathrm{b}}\mathbf{v}_i = {}^{\mathrm{g}}_{\mathrm{b}}\mathrm{R}\,{}_{\mathrm{g}}\mathbf{v}_i, \tag{9.1}
$$

where ${}^{\mathrm{b}}_{\mathrm{g}}\mathrm{R} = {}^{\mathrm{g}}_{\mathrm{b}}\mathrm{R}^{-1}$ and $i = 1, \cdots, N$ is the sensor index. People have formulated the problem from the perspective of optimization, which aims to minimize the following cost function, known as Wahba's problem

$$
J({}^{\mathrm{g}}_{\mathrm{b}}\mathrm{R}) = \frac{1}{2}\sum_{i=1}^{N} w_i \|{}_{\mathrm{b}}\mathbf{v}_i - {}^{\mathrm{g}}_{\mathrm{b}}\mathrm{R}\,{}_{\mathrm{g}}\mathbf{v}_i\|^2 \tag{9.2}
$$

where w_i denotes the weight for the ith sensor. Figure 9.1 illustrates one example, where two sensors are used to measure one ambient vector field.

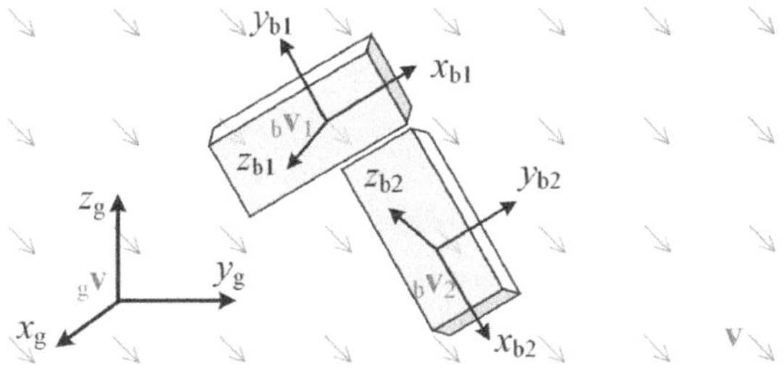

Figure 9.1 Heading reference-based attitude estimation with two heading reference sensors measuring a unique ambient vector field **v**.

Various methods can be used to optimize $J({}_{\mathrm{b}}^{\mathrm{g}}\mathrm{R})$. QUEST, which estimates rotation in parameterization $\mathbf{q} = [q_x, q_y, q_z, q_w]^{\mathrm{T}}$, is one of the most popular

$$\mathbf{q} = \frac{1}{\sqrt{1+\mathbf{p}^{\mathrm{T}}\mathbf{p}}}\begin{bmatrix}\mathbf{p}\\1\end{bmatrix}, \tag{9.3}$$

where $\mathbf{p} = [(\lambda_{\mathrm{opt}} + \sigma_{\mathrm{q}})\mathbf{I} - \mathbf{S}_{\mathrm{q}}]^{-1}\mathbf{Z}_{\mathrm{q}}$, and

$$\sigma_{\mathrm{q}} = \mathrm{trace}(\mathbf{B}_{\mathrm{q}}),\ \mathbf{S}_{\mathrm{q}} = \mathbf{B}_{\mathrm{q}} + \mathbf{B}_{\mathrm{q}}^{\mathrm{T}},\ \mathbf{B}_{\mathrm{q}} = \sum_i w_i \left({}_{\mathrm{G}}\mathbf{v}_{\mathrm{B}}\mathbf{v}^{\mathrm{T}}\right),$$

$$\mathbf{Z}_{\mathrm{q}} = \sum_i w_i\, {}_{\mathrm{b}}\mathbf{v}_i \times {}_{\mathrm{g}}\mathbf{v}_i,\ \mathbf{L}_{\mathrm{q}} = \begin{bmatrix}\mathbf{S}_{\mathrm{q}} - \sigma_{\mathrm{q}}\mathbf{I} & \mathbf{Z}_{\mathrm{q}}\\ \mathbf{Z}_{\mathrm{q}}^{\mathrm{T}} & \sigma_{\mathrm{q}}\end{bmatrix}.$$

The identity matrix is denoted by **I**, and λ_{opt} is the maximum eigenvalue of $\mathbf{L}_{\mathrm{q}}$, which can be obtained by the Newton-Raphson method at initial estimation $\lambda_{\mathrm{opt0}} = \sum_i w_i$.

Various types of sensors can be used as heading reference sensors. Magnetometers and accelerometers are popular heading reference sensors that detect linear accelerations and magnetic fields in three independent (and perpendicular) directions. In spacecraft, star sensors and directional antennas have also been used to acquire accurate pose estimation without drift error.

9.3.2 Secure Attitude Estimation Framework and Modeling

Secure estimation can be achieved on a lower level, targeting raw measurements, or on a higher level for estimated attitude. In this chapter, we propose to implement a secure filter on raw vector field measurements; cost function minimization can then be done by following the conventional method. The proposed framework is illustrated in Figure 9.2, in which the secure filter serves as the core contribution.

We consider the system with n heading reference sensors. Since ${}_{\mathrm{g}}\mathbf{y}_i$ is usually a priori, which is not deemed a measurement, by selecting raw measurements as states, the system state vector can be represented as $\mathbf{x} = \begin{bmatrix}{}_{\mathrm{b}}\mathbf{v}_1 & \cdots & {}_{\mathrm{b}}\mathbf{v}_n\end{bmatrix}^{\mathrm{T}} \in \mathbb{R}^{3n}$, where ${}_{\mathrm{b}}\mathbf{v}_i \in \mathbb{R}^3$ denotes a measurement from a single sensor.

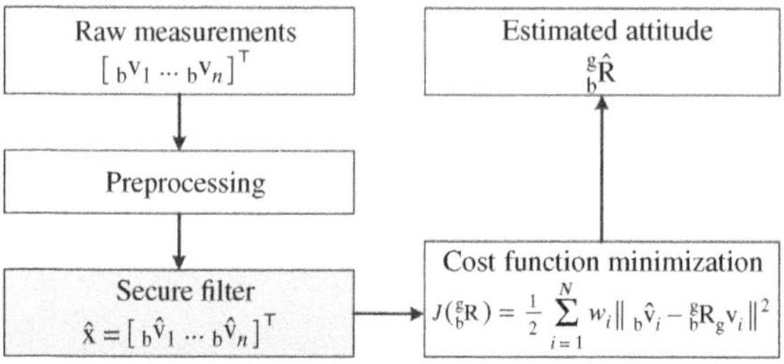

Figure 9.2 Secure attitude estimation framework. This chapter focuses on designing a secure filter for state estimation.

As we have no a priori information about attitude, the process is modeled as constant with Gaussian-distributed noises based on the low dynamics assumption [Kelly, 2013]. The following linear time-invariant system is used

$$\mathbf{x}(k+1) = \mathbf{A}\mathbf{x}(k) + \mathbf{w}(k) \tag{9.4}$$

$$\mathbf{y}(k) = \mathbf{C}\mathbf{x}(k) + \mathbf{v}(k) + \mathbf{a}(k) = \mathbf{z}(k) + \mathbf{a}(k) \tag{9.5}$$

where $\mathbf{a}(k)$ denotes the bias vector injected by an adversary, $\mathbf{z}(k)$ is regarded as the true measurement vector without attack, matrices $\mathbf{A} = \mathbf{C} = \mathbf{I}$ are both identity matrices, and $\mathbf{w}(k) \sim \mathcal{N}(0, \mathbf{Q})$, $\mathbf{v}(k) \sim \mathcal{N}(0, \mathbf{R})$ are independent and identically distributed (i.i.d.) and mutually independent Gaussian process and measurement noises, respectively.

Remark 9.1 It should be mentioned that there is no bound limitation for the sparse attack $\mathbf{a}(k)$. Although the system is subject to sector-bounded non-linearity, unknown but bounded noises and sensor saturations have been investigated in Ma et al. (2017b); the sparse attack $\mathbf{a}(k)$ in this chapter cannot be a bound, and the similar treatments of Ma et al. (2017b) cannot be applicable. Hence, this chapter investigates the problem of secure attitude estimation for the AHRSs measuring model under sparse attacks.

9.4 Secure Estimator Design with a Decomposing Kalman Filter

Throughout this chapter, we define the following notations:

$$\mathbf{y}(k) \triangleq \begin{bmatrix} \mathbf{y}_1(k) \\ \vdots \\ \mathbf{y}_{3n}(k) \end{bmatrix}, \ \mathbf{z}(k) \triangleq \begin{bmatrix} \mathbf{z}_1(k) \\ \vdots \\ \mathbf{z}_{3n}(k) \end{bmatrix}, \ \mathbf{C} \triangleq \begin{bmatrix} \mathbf{C}_1 \\ \vdots \\ \mathbf{C}_{3n} \end{bmatrix},$$
$$\mathbf{a}(k) \triangleq \begin{bmatrix} \mathbf{a}_1(k) \\ \vdots \\ \mathbf{a}_{3n}(k) \end{bmatrix}, \ \mathbf{v}(k) \triangleq \begin{bmatrix} \mathbf{v}_1(k) \\ \vdots \\ \mathbf{v}_{3n}(k) \end{bmatrix}, \tag{9.6}$$

where $\mathbf{y}_i(k) \in \mathbb{R}$, $\mathbf{z}_i(k) \in \mathbb{R}$, $\mathbf{a}_i(k) \in \mathbb{R}$, $\mathbf{v}_i(k) \in \mathbb{R}$, and $\mathbf{C}_i = [\underbrace{0 \cdots 0}_{i-1} 1\ 0 \cdots 0] \in \mathbb{R}^{1\times 3n}$ for $i = 1, 2, \cdots, 3n$.

If all the elements of the model state are benign, i.e. $\mathbf{a}(k) = 0$ for all k, the optimal state estimator is the classical Kalman filter

$$\hat{\mathbf{x}}(k) = \hat{\mathbf{x}}(k|k-1) + \mathbf{K}(k)\left[\mathbf{y}(k) - \hat{\mathbf{x}}(k|k-1)\right],$$
$$\mathbf{P}(k) = \mathbf{P}(k|k-1) - \mathbf{K}(k)\mathbf{P}(k|k-1),$$

where

$$\hat{\mathbf{x}}(k+1|k) = \hat{\mathbf{x}}(k),\ \ \mathbf{P}(k+1|k) = \mathbf{P}(k) + \mathbf{Q},$$
$$\mathbf{K}(k) = \mathbf{P}(k|k-1)(\mathbf{P}(k|k-1) + \mathbf{R})^{-1},$$

with initial condition $\hat{\mathbf{x}}(0|-1) = 0$, $\mathbf{P}(0|-1) = \boldsymbol{\Sigma}$.

Since the system is observable, it is well known that the estimation error covariance matrices $\mathbf{P}(k)$ and the gain $\mathbf{K}(k)$ will converge to

$$\mathbf{P} \triangleq \lim_{k\to\infty} \mathbf{P}(k),\ \ \mathbf{P}_+ = \mathbf{P} + \mathbf{Q},\ \ \mathbf{K} \triangleq \mathbf{P}_+(\mathbf{P}_+ + \mathbf{R})^{-1}.$$

We assume that the Kalman filter is in a steady state or, equivalently, that $\boldsymbol{\Sigma} = \mathbf{P}$, and thus the Kalman filter reduces to the following fixed-gain linear estimator:

$$\hat{\mathbf{x}}(k+1) = (\mathbf{I} - \mathbf{K})\hat{\mathbf{x}}(k) + \mathbf{K}\mathbf{y}(k+1). \tag{9.7}$$

Define the matrix $\mathbf{K} = [\mathbf{K}_1, \dots, \mathbf{K}_{3n}]$. Accordingly, (9.7) can be rewritten as

$$\hat{\mathbf{x}}(k+1) = (\mathbf{I} - \mathbf{K})\hat{\mathbf{x}}(k) + \sum_{i=1}^{3n} \mathbf{K}_i\mathbf{y}_i(k+1). \tag{9.8}$$

We denote the index set of all state elements as $\mathcal{S} = \{1, 2, \cdots, 3n\}$. For any index set $\mathcal{I} \subseteq \mathcal{S}$, define the complement set to be $\mathcal{I}^c = \mathcal{S}\backslash\mathcal{I}$. For the attack model discussed in this chapter, we assume that the attacker can only compromise at most p elements but can arbitrarily choose $\mathbf{a}_i$. Furthermore, we define the collection of all possible index sets of malicious elements as $\mathcal{C} \triangleq \{\mathcal{I} : \mathcal{I} \subseteq \mathcal{S}, |\mathcal{I}| = p\}$. The set of all possible $(p, 3n)$-sparse attacks is denoted as $\mathcal{A} \triangleq \bigcup_{\mathcal{I}\in\mathcal{C}}\{\mathbf{a}(k) : \ \|\mathbf{a}_i(k)\| = 0,\ i \in \mathcal{I}^c\}$ for all k.

The objective of this chapter is to design a framework that can estimate the state of the AHRS measuring system in such a way that:

1. If no element of the system state is compromised, i.e. $\mathbf{a}_i(k) = 0$ for all i and all k, the estimate coincides with a certain probability to the standard Kalman estimate.
2. We establish a condition under which the estimator is stable against a $(p, 3n)$-sparse attack, i.e. the estimator has a bounded estimation error when p elements of the model state are compromised.

9.4.1 Decomposition of the Kalman Filter Using a Local Estimate

In this subsection, we introduce a method to decompose the Kalman estimate (9.7) into a linear combination of local state estimates. We will make the following assumption throughout this chapter.

Assumption [Liu et al., 2017] $\mathbf{I}-\mathbf{K}$ can be diagonalized. Moreover, $\mathbf{I}-\mathbf{K}$ and $\mathbf{I}$ do not share any eigenvalues. □

$\mathbf{I}-\mathbf{K}$ can be diagonalized as

$$\mathbf{I}-\mathbf{K}=\mathbf{V}\mathbf{\Lambda}\mathbf{V}^{-1}. \tag{9.9}$$

Subsequently, we can rewrite (9.8) as

$$\begin{aligned}\left[\mathbf{V}^{-1}\hat{\mathbf{x}}(k+1)\right] &= \mathbf{\Lambda}\left[\mathbf{V}^{-1}\hat{\mathbf{x}}(k)\right]\\ &\quad+\sum_{i=1}^{3n}\mathbf{V}^{-1}\mathbf{K}_i\mathbf{y}_i(k+1).\end{aligned} \tag{9.10}$$

Our goal is to generate a local state estimate $\hat{\zeta}_i(k)$, $i=1,\dots,3n$ such that

$$\hat{\mathbf{x}}(k)=\mathbf{F}_1\hat{\zeta}_1(k)+\cdots+\mathbf{F}_{3n}\hat{\zeta}_{3n}(k).$$

Now consider the following recursive equation

$$\hat{\zeta}_i(k+1)=\mathbf{\Lambda}\hat{\zeta}_i(k)+\mathbf{1}_{3n}\mathbf{y}_i(k+1), \tag{9.11}$$

where $\mathbf{1}_{3n}\in\mathbb{R}^{3n\times 1}$ is an all-one vector and $\mathbf{\Lambda}$ is defined in (9.9).

Let us choose $\mathbf{F}_i$ as $\mathbf{F}_i=\mathbf{V}diag(\mathbf{V}^{-1}\mathbf{K}_i)$, where $\mathbf{V}$ is defined in (9.9) and $diag(\mathbf{V}^{-1}\mathbf{K}_i)$ is an $3n\times 3n$ diagonal matrix with the jth diagonal entry equal to the jth entry of the vector $\mathbf{V}^{-1}\mathbf{K}_i$. Comparing (9.10) and (9.11), we can obtain that

$$\hat{\mathbf{x}}(k)=\sum_{i=1}^{3n}\mathbf{F}_i\hat{\zeta}_i(k). \tag{9.12}$$

Define the matrix $\mathbf{H}_i$ as

$$\mathbf{H}_i\triangleq\begin{bmatrix}(1-\lambda_1)^{-1}\mathbf{C}_i\\ \vdots\\ (1-\lambda_{3n})^{-1}\mathbf{C}_i\end{bmatrix}\in\mathbb{R}^{3n\times 3n},$$

where $i=1,2,\cdots,3n$ and λ_i is the ith eigenvalues of $\mathbf{I}-\mathbf{K}$.

Let $\epsilon_i(k)\triangleq\mathbf{H}_i\mathbf{x}(k)-\hat{\zeta}_i(k)$. Then we obtain that

$$\begin{aligned}\epsilon_i(k+1)&=\mathbf{H}_i\mathbf{x}(k+1)-\hat{\zeta}_i(k+1)\\ &=(\mathbf{H}_i-\mathbf{1}_{3n}\mathbf{C}_i)\mathbf{x}(k)-\mathbf{\Lambda}\hat{\zeta}_i(k)-\mathbf{1}_{3n}\mathbf{a}_i(k+1)\\ &\quad+(\mathbf{H}_i-\mathbf{1}_{3n}\mathbf{C}_i)\mathbf{w}(k)-\mathbf{1}_{3n}\mathbf{v}_i(k+1).\end{aligned}$$

By the definition of $\mathbf{H}_i$, it can be verified that

$$\mathbf{H}_i - \mathbf{1}_{3n}\mathbf{C}_i = \begin{bmatrix} (1-\lambda_1)^{-1}\mathbf{C}_i - \mathbf{C}_i \\ \vdots \\ (1-\lambda_{3n})^{-1}\mathbf{C}_i - \mathbf{C}_i \end{bmatrix} = \Lambda \mathbf{H}_i. \tag{9.13}$$

Therefore,

$$\begin{aligned} \epsilon_i(k+1) &= \Lambda \mathbf{H}_i \mathbf{x}(k) - \Lambda \hat{\zeta}_i(k) - \mathbf{1}_{3n}\mathbf{a}_i(k+1) \\ &\quad + (\mathbf{H}_i - \mathbf{1}_{3n}\mathbf{C}_i)\mathbf{w}(k) - \mathbf{1}_{3n}\mathbf{v}_i(k+1) \\ &= \Lambda \epsilon_i(k) + (\mathbf{H}_i - \mathbf{1}_{3n}\mathbf{C}_i)\mathbf{w}(k) \\ &\quad - \mathbf{1}_{3n}\mathbf{v}_i(k+1) - \mathbf{1}_{3n}\mathbf{a}_i(k+1). \end{aligned} \tag{9.14}$$

It can be seen that $\hat{\zeta}_i(k)$ is a stable estimate of $\mathbf{H}_i\mathbf{x}(k)$ since $\mathbf{I} - \mathbf{K}$ is stable.

9.4.2 A Least-Square Interpretation for the Decomposition

In this subsection, we show that the decomposition (9.12) can be interpreted as a least-square problem, which will be used later to derive a secure fusion scheme.

Based on the recursive equation (9.14), we let $\epsilon_i(k) = \boldsymbol{\mu}_i(k) + \boldsymbol{\nu}_i(k)$ and choose as

$$\begin{aligned} \boldsymbol{\mu}_i(k+1) = \Lambda \boldsymbol{\mu}_i(k) + (\mathbf{H}_i - \mathbf{1}_{3n}\mathbf{C}_i)\mathbf{w}(k) \\ -\mathbf{1}_{3n}\mathbf{v}_i(k+1), \end{aligned} \tag{9.15}$$

$$\boldsymbol{\nu}_i(k+1) = \Lambda \boldsymbol{\nu}_i(k) - \mathbf{1}_{3n}\mathbf{a}_i(k+1). \tag{9.16}$$

where $\boldsymbol{\mu}_i(k)$ can be regarded as the error of the local estimate caused by noise and $\boldsymbol{\nu}_i(k)$ as the error caused by the bias injected by the adversary.

Define $\tilde{\Lambda} \in \mathbb{R}^{9n^2 \times 9n^2}$, $\tilde{\boldsymbol{\mu}}(k) \in \mathbb{R}^{9n^2}$ as

$$\tilde{\Lambda} \triangleq \begin{bmatrix} \Lambda & & \\ & \ddots & \\ & & \Lambda \end{bmatrix}, \quad \tilde{\boldsymbol{\mu}}(k) \triangleq \begin{bmatrix} \boldsymbol{\mu}_1(k) \\ \vdots \\ \boldsymbol{\mu}_{3n}(k) \end{bmatrix}. \tag{9.17}$$

Similarly, $\tilde{\epsilon}(k)$ and $\tilde{\nu}(k)$ are defined by stacking $\epsilon_i(k)$ and $\boldsymbol{\nu}_i(k)$. Taking account of (9.15), we have that

$$Cov[\tilde{\boldsymbol{\mu}}(k+1)] = \tilde{\Lambda} Cov[\tilde{\boldsymbol{\mu}}(k)]\tilde{\Lambda}^{\top} + \tilde{\mathbf{Q}} \tag{9.18}$$

where

$$\begin{aligned} \tilde{\mathbf{Q}} &\triangleq Cov\left[\begin{pmatrix} \mathbf{H}_1 - \mathbf{1}_{3n}\mathbf{C}_1 \\ \vdots \\ \mathbf{H}_{3n} - \mathbf{1}_{3n}\mathbf{C}_{3n} \end{pmatrix}\mathbf{w}(k)\right] \\ &\quad + Cov\left[\begin{pmatrix} \mathbf{1}_{3n}\mathbf{v}_1(k+1) \\ \vdots \\ \mathbf{1}_{3n}\mathbf{v}_{3n}(k+1) \end{pmatrix}\right], \end{aligned} \tag{9.19}$$

and the covariances are described as

$$
\begin{aligned}
&Cov\left[\begin{pmatrix} \mathbf{H}_1 - \mathbf{1}_{3n}\mathbf{C}_1 \\ \vdots \\ \mathbf{H}_{3n} - \mathbf{1}_{3n}\mathbf{C}_{3n} \end{pmatrix}\mathbf{w}(k)\right] \\
&= \begin{bmatrix} \mathbf{H}_1 - \mathbf{1}_{3n}\mathbf{C}_1 \\ \vdots \\ \mathbf{H}_{3n} - \mathbf{1}_{3n}\mathbf{C}_{3n} \end{bmatrix} \mathbf{Q} \begin{bmatrix} \mathbf{H}_1 - \mathbf{1}_{3n}\mathbf{C}_1 \\ \vdots \\ \mathbf{H}_{3n} - \mathbf{1}_{3n}\mathbf{C}_{3n} \end{bmatrix}^{\mathsf{T}}, \qquad (9.20)
\end{aligned}
$$

$$
Cov\left[\begin{pmatrix} \mathbf{1}_{3n}\mathbf{v}_1(k+1) \\ \vdots \\ \mathbf{1}_{3n}\mathbf{v}_{3n}(k+1) \end{pmatrix}\right] = \mathbf{1}_{9n^2 \times 9n^2} \circ \left(\mathbf{R} \otimes \mathbf{1}_{3n \times 3n}\right), \qquad (9.21)
$$

where $\circ$ denotes element-wise matrix multiplication and $\otimes$ is the Kronecker product.

Now we introduce the following least-square optimization problem:

$$
\begin{aligned}
\underset{\check{\mathbf{x}}(k),\check{\epsilon}(k)}{\text{minimize}} \quad & \frac{1}{2}\check{\epsilon}(k)^{\mathsf{T}}\tilde{\mathbf{W}}^{-1}\check{\epsilon}(k) \qquad (9.22) \\
\text{subject to} \quad & \begin{bmatrix} \hat{\zeta}_1(k) \\ \vdots \\ \hat{\zeta}_{3n}(k) \end{bmatrix} = \begin{bmatrix} \mathbf{H}_1 \\ \vdots \\ \mathbf{H}_{3n} \end{bmatrix} \check{\mathbf{x}}(k) - \check{\epsilon}(k).
\end{aligned}
$$

where $\tilde{\mathbf{W}} = \tilde{\mathbf{\Lambda}}\tilde{\mathbf{W}}\tilde{\mathbf{\Lambda}}^{\mathsf{T}} + \tilde{\mathbf{Q}}$.

Remark 9.2 This can be interpreted as the problem of finding an estimate $\check{\mathbf{x}}(k)$ that minimizes a weighted least square of the error with the local estimates $\hat{\zeta}_i(k)$, where the weighting matrix is related with the covariance of the error of the local estimates.

It should be noted that the least-square problem (9.22) is not secure in the sense that if the ith element of the state is compromised, the adversary can manipulate $\hat{\zeta}_i(k)$ by injecting the bias $\mathbf{a}_i(k)$ into the measurement $\mathbf{y}_i(k)$. Hence, the adversary can potentially change the Kalman estimate arbitrarily.

9.4.3 Secure State Estimate

In this subsection, we introduce a convex optimization-based approach to combine the local estimate into a more secure state estimate; a secure state estimate is then achieved by solving the convex optimization problem.

As explained earlier, $\epsilon_i(k)$ can be decomposed as the error caused by the noise $\mu_i(k)$ and the error caused by the bias injected by the adversary $\nu_i(k)$. Motivated by this observation, the secure fusion scheme is proposed as follows

$$
\begin{aligned}
\underset{\check{\mathbf{x}}_s(k),\check{\mu}(k),\check{\nu}(k)}{\text{minimize}} \quad & \frac{1}{2}\check{\mu}(k)^{\mathsf{T}}\tilde{\mathbf{W}}^{-1}\check{\mu}(k) + \gamma\|\check{\nu}(k)\|_1 \qquad (9.23) \\
\text{subject to} \quad & \hat{\zeta}_i(k) = \mathbf{H}_i\check{\mathbf{x}}_s(k) - \check{\mu}_i(k) - \check{\nu}_i(k), \ \forall i \in \mathcal{S},
\end{aligned}
$$

where $\check{\mathbf{x}}_s(k)$ is the secure state estimation, γ is a constant chosen by the system operator, and $\check{\mu}(k)$ and $\check{\nu}(k)$ are defined as

$$\check{\mu}(k) \triangleq \begin{bmatrix} \check{\mu}_1(k) \\ \vdots \\ \check{\mu}_{3n}(k) \end{bmatrix}, \quad \check{\nu}(k) \triangleq \begin{bmatrix} \check{\nu}_1(k) \\ \vdots \\ \check{\nu}_{3n}(k) \end{bmatrix}.$$

We now consider two scenarios:

1. There is no compromised element of the system state, i.e. $\mathbf{a}(k) = 0$ for all k, and the system is operating normally.
2. $p(p < \lfloor \frac{3n}{2} \rfloor)$ elements of the system state are compromised, where $\lfloor x \rfloor$ is a floor function, i.e. $\lfloor x \rfloor = \max\{m \in \mathbb{Z} : m \leq x\}$.

Corollary 9.1 *Let* $\check{\mathbf{x}}_s(k)$, $\check{\mu}(k)$, $\check{\nu}(k)$ *be the minimizer for the optimization problem* (9.23). *Let* $\check{\mathbf{x}}(k)$, $\check{e}(k)$ *be the minimizer for the least-square problem* (9.22). *Then the following statements hold:*

1. *The following inequality holds:*

$$\| \tilde{\mathbf{W}}^{-1} \check{\mu}(k) \|_\infty \leq \gamma. \tag{9.24}$$

2. *For the first scenario, we conclude that* $\check{\mathbf{x}}_s(k) = \check{\mathbf{x}}(k) = \hat{\mathbf{x}}(k)$ *if the following inequality holds:*

$$\left\| \tilde{\mathbf{W}}^{-1} \left(\mathbf{I} - \begin{bmatrix} \mathbf{H}_1 \\ \vdots \\ \mathbf{H}_{3n} \end{bmatrix} \begin{bmatrix} \mathbf{F}_1 & \dots & \mathbf{F}_{3n} \end{bmatrix} \right) \check{e}(k) \right\|_\infty \leq \gamma. \tag{9.25}$$

Proof: A similar result and detailed procedures can be referred to in Liu et al. (2017). Hence the proof is omitted. □

For the second scenario, we first present a lemma, and the stability of the proposed secure estimator is then characterized by Theorem 9.1.

Lemma 9.1 *Considering the following optimization problem*

$$\mathbf{e} = \arg\min \sum_{i=1}^{3n} \| \boldsymbol{\eta}_i - \mathbf{H}_i \mathbf{e} \|_1 = \mathbf{g}(\boldsymbol{\eta}_1, \dots, \boldsymbol{\eta}_{3n}), \tag{9.26}$$

where $\mathbf{e}$ *is the optimal solution of the optimization problem and* $\boldsymbol{\eta}_1, \dots, \boldsymbol{\eta}_{3n}$ *are independent variables of the vector function* $\mathbf{g}(\cdot)$. *If the following inequality holds for all* $\mathbf{u} \neq 0$

$$\sum_{i \in \mathcal{I}} \| \mathbf{H}_i \mathbf{u} \|_1 < \sum_{i \in \mathcal{I}^c} \| \mathbf{H}_i \mathbf{u} \|_1, \ \forall \mathcal{I} \in C,$$

then the following statements hold:

1. *For* $i = 1, 2, \cdots, 3n$*, we have*

$$\|\mathbf{e}\|_1 \leq \frac{6n \max_i \|\boldsymbol{\eta}_i\|_1}{\delta_1}. \tag{9.27}$$

2. *Assuming that* $\mathbf{e}' = \mathbf{g}(\boldsymbol{\eta}'_1, \dots, \boldsymbol{\eta}'_{3n})$*, where* $\boldsymbol{\eta}_i - \boldsymbol{\eta}'_i = \mathbf{v}_i$ *and at most* p $\mathbf{v}_i$*s are non-zero, then* $\|\mathbf{e} - \mathbf{e}'\|_1$ *is bounded regardless of* $\mathbf{v}_1, \dots, \mathbf{v}_{3n}$*, i.e.*

$$\|\mathbf{e} - \mathbf{e}'\|_1 \leq \frac{2\max_i \| \boldsymbol{\eta}'_i\|_1 + \delta_2}{\delta_2} + \frac{6n \max_i \| \boldsymbol{\eta}'_i\|_1}{\delta_1}, \tag{9.28}$$

where $\delta_1 = \min_{\|\mathbf{u}\|_1=1} \sum_{i=1}^{3n} \|\mathbf{H}_i\mathbf{u}\|_1$ *and*

$$\delta_2 = \frac{1}{3n} \min_{\|\mathbf{u}\|_1=1} \min_{\mathcal{I}\in C} \left(\sum_{i\in\mathcal{I}^c} \|\mathbf{H}_i\mathbf{u}\|_1 - \sum_{i\in\mathcal{I}} \|\mathbf{H}_i\mathbf{u}\|_1 \right). \tag{9.29}$$

Proof: The proof is omitted here. Interested readers can refer to Han et al. (2015), where a similar proof can be found. □

Theorem 9.1 *Suppose that* $p (p < \lfloor \frac{3n}{2} \rfloor)$ *elements of the system state are compromised. Then the secure state estimate* $\check{\mathbf{x}}_s(k)$ *is stable against a* $(p, 3n)$*-sparse attack if the following inequality holds for all* $\mathbf{u} \neq 0$*:*

$$\sum_{i\in\mathcal{I}} \|\mathbf{H}_i\mathbf{u}\|_1 < \sum_{i\in\mathcal{I}^c} \|\mathbf{H}_i\mathbf{u}\|_1, \ \forall \mathcal{I} \in C. \tag{9.30}$$

Proof: Let us define $\mathbf{e}_s(k) = \mathbf{x}(k) - \check{\mathbf{x}}_s(k)$. By subtracting $\mathbf{H}_i\mathbf{x}$ on both sides of the constraint equation in optimization problem (9.23), it is easy to prove that $\mathbf{e}_s(k)$ is the solution of the following minimization problem:

$$\begin{aligned} &\underset{\mathbf{e}_s(k),\check{\boldsymbol{\mu}}(k),\check{\mathbf{v}}(k)}{\text{minimize}} && \frac{1}{2}\check{\boldsymbol{\mu}}(k)^{\mathrm{T}}\tilde{\mathbf{W}}^{-1}\check{\boldsymbol{\mu}}(k) + \gamma\|\check{\mathbf{v}}(k)\|_1 \\ &\text{subject to} && \mu_i(k) + v_i(k) = \mathbf{H}_i\mathbf{e}_s(k) + \check{\mu}_i(k) + \check{\mathbf{v}}_i(k). \end{aligned} \tag{9.31}$$

Let us define

$$\boldsymbol{\eta}_i(k) = \boldsymbol{\mu}_i(k) + \mathbf{v}_i(k) - \check{\boldsymbol{\mu}}_i(k), \ \ \boldsymbol{\eta}'_i(k) = \boldsymbol{\mu}_i(k) - \check{\boldsymbol{\mu}}_i(k). \tag{9.32}$$

According to the optimization problem (9.26), apparently we know that $\mathbf{e}_s(k) = \mathbf{g}(\boldsymbol{\eta}_1, \dots, \boldsymbol{\eta}_{3n})$.

By (9.28) in Lemma 9.1, for all k, we obtain that

$$\begin{aligned} &\| \mathbf{e}_s(k) - \mathbf{g}(\boldsymbol{\eta}'_1(k), \dots, \boldsymbol{\eta}'_{3n}(k))\|_1 \\ &\leq \frac{2\max_i\|\boldsymbol{\eta}'_i(k)\|_1 + \delta_2}{\delta_2} + \frac{6n \max_i\|\boldsymbol{\eta}'_i(k)\|_1}{\delta_1}. \end{aligned}$$

Furthermore, by (9.27) in Lemma 9.1, for all k, we have

$$\|\mathbf{g}(\eta_1'(k), \dots, \eta_{3n}'(k))\|_1 \leq \frac{6n \max_i \|\eta_i'(k)\|_1}{\delta_1}.$$

Therefore, it implies that

$$\begin{aligned}
&\|\mathbf{x}(k) - \check{\mathbf{x}}_s(k)\|_1 = \|\mathbf{e}_s(k)\|_1 \\
&\leq \|\mathbf{e}_s(k) - \mathbf{g}(\eta_1'(k), \dots, \eta_{3n}'(k))\|_1 + \|\mathbf{g}(\eta_1'(k), \dots, \eta_{3n}'(k))\|_1 \\
&\leq \frac{2\max_i \|\eta_i'(k)\|_1 + \delta_2}{\delta_2} + \frac{12n \max_i \|\eta_i'(k)\|_1}{\delta_1}. \qquad (9.33)
\end{aligned}$$

According to the inequality (9.24), we know that $\check{\mu}_i(k)$ is bounded. So we can claim that the bound of the estimation error is irrelevant to the attack.

Since the estimation error is bounded as (9.33), we can conclude the secure estimator $\check{\mathbf{x}}_s(k)$ is stable against the $(p, 3n)$-sparse attack. This completes the proof. □

For clarity, we summarize the proposed secure state estimate procedure in Algorithm 6.

Algorithm 1 Secure Estimate by Decomposing the Kalman Filter

Step 1. For given $\mathbf{A} = \mathbf{C} = \mathbf{I}$, $\mathbf{w}(k) \sim \mathcal{N}(0, \mathbf{Q})$,$\mathbf{v}(k) \sim \mathcal{N}(0, \mathbf{R})$, solve the Kalman fixed-gain $\mathbf{K}$.

Step 2. Diagonalize the matrix $\mathbf{I} - \mathbf{K}$, and obtain $\mathbf{\Lambda}$, $\mathbf{H}_i$. Then let $\epsilon_i(k) = \boldsymbol{\mu}_i(k) + \boldsymbol{\nu}_i(k)$ and

$$\begin{aligned}
\boldsymbol{\mu}_i(k+1) &= \mathbf{\Lambda}\boldsymbol{\mu}_i(k) + (\mathbf{H}_i - \mathbf{1}_{3n}\mathbf{C}_i)\mathbf{w}(k) - \mathbf{1}_{3n}\mathbf{v}_i(k+1), \\
\boldsymbol{\nu}_i(k+1) &= \mathbf{\Lambda}\boldsymbol{\nu}_i(k) - \mathbf{1}_{3n}\mathbf{a}_i(k+1).
\end{aligned}$$

Step 3. Set the iterations, i.e. the value of T. For k=0, solve the optimization problem as follows:

$$\begin{aligned}
&\underset{\check{\mathbf{x}}_s(k), \check{\mu}(k), \check{\nu}(k)}{\text{minimize}} && \frac{1}{2}\check{\mu}(k)^\top \tilde{\mathbf{W}}^{-1}\check{\mu}(k) + \gamma\|\check{\nu}(k)\|_1 \\
&\text{subject to} && \hat{\zeta}_i(k) = \mathbf{H}_i\check{\mathbf{x}}_s(k) - \check{\mu}_i(k) - \check{\nu}_i(k), \ \forall i \in \mathcal{S}.
\end{aligned}$$

Step 4. Set k=k+1, and go to Step 3; when $k = T$, stop.

9.5 Simulation Validation

Without loss of generality, we consider the problem with two heading reference sensors ($n = 2$) in all experiments. Based on the assumption that components in heading reference sensors are mutually independent, the covariance matrices of noises are set as

$$\mathbf{Q} = \text{diag}(3, 1, 4, 5, 2, 6) \times 10^{-4}, \ \mathbf{R} = 0.01\mathbf{I}_6.$$

To solve the optimization problem, we used CVX, a package for specifying and solving convex programs (http://cvxr.com/cvx).

9.5.1 Simulating Measurements with Attacks

All attacks are intentionally generated, which makes them hard to test with real experiments. So, to better evaluate estimator performance with respect to different γ, artificially designed attack signals have been used in the simulations. To compare the estimated results and ground truth, all measurements are simulated based on the inverse procedure of state estimation. Suppose we have predetermined ambient vectors in the global frame

$$ {}_{\mathrm{g}}\mathbf{v}_1 = \begin{bmatrix} 1 & 1 & 1 \end{bmatrix}^{\mathrm{T}}, \quad {}_{\mathrm{g}}\mathbf{v}_2 = \begin{bmatrix} \cos\gamma & -\sin\gamma & 0 \\ \sin\gamma & \cos\gamma & 0 \\ 0 & 0 & 1 \end{bmatrix} {}_{\mathrm{g}}\mathbf{v}_1, $$

where $\gamma = \frac{\pi}{2}$. Without loss of generality, we let $\mathbf{q}_0 = \left[0.462, 0.191, 0.462, 0.733\right]^{\mathrm{T}}$, which indicates a random pose; and ${}^{\mathrm{g}}_{\mathrm{b}}\mathrm{R}(0) = \mathtt{quat2rotm}(\mathbf{q}_0)$, where $\mathtt{quat2rotm}$ $(\mathbf{q}_0)$ converts the unit quaternion $\mathbf{q}_0$ to a rotation matrix representation. With a rotation matrix ${}^{\mathrm{g}}_{\mathrm{b}}\mathrm{R}$ designed, the measurement vectors in the body frame ${}_{\mathrm{b}}\mathbf{v}_1$ and ${}_{\mathrm{b}}\mathbf{v}_2$ can be obtained immediately from (9.1). The attack models in the simulations are $\mathbf{a}_1(k) = \left[100, \mathbf{0}_{1\times5}\right]^{\mathrm{T}}$, and $\mathbf{a}_2(k) = \left[100, \mathbf{0}_{1\times2}, 100, \mathbf{0}_{1\times2}\right]^{\mathrm{T}}$ for all k, as examples for the single-state attack and multiple states attack, respectively.[1] The discrete time steps are set to 50 in all simulations.

9.5.2 Filter Performance

The simulation results are shown in Figure 9.3 and Figure 9.4, where Figure 9.3a and 9.4a illustrate the state estimation error without any attack for $\gamma = 0.0001$ and $\gamma = 5000$, respectively. We observe that all estimated states tend to approach true states within an error upper bound, and $\gamma = 0.0001$ leads to faster convergence than $\gamma = 5000$. This phenomenon can be explained based on the previous discussion that a small γ makes the optimization problem consider less influence from the error caused by the bias injected by the adversary.

The state estimation errors with attacks are shown in Figure 9.3b, Figure 9.3c, Figure 9.4b, and Figure 9.4c. Similar to the non-attacked results, we see that the estimation error of non-attacked states is bounded. The attacked states are inevitably influenced, while the filter with $\gamma = 5000$ gives a lower mean squared error (MSE) than the results with $\gamma = 0.0001$.

9.5.3 Influence of Parameter γ

To further investigate the estimator's performance with regard to γ, we compare the MSE of the proposed secure filter and the Kalman filter. The values of their

1 The sparse attack assumption requires us to apply attacks on fewer than three measurement vector components.

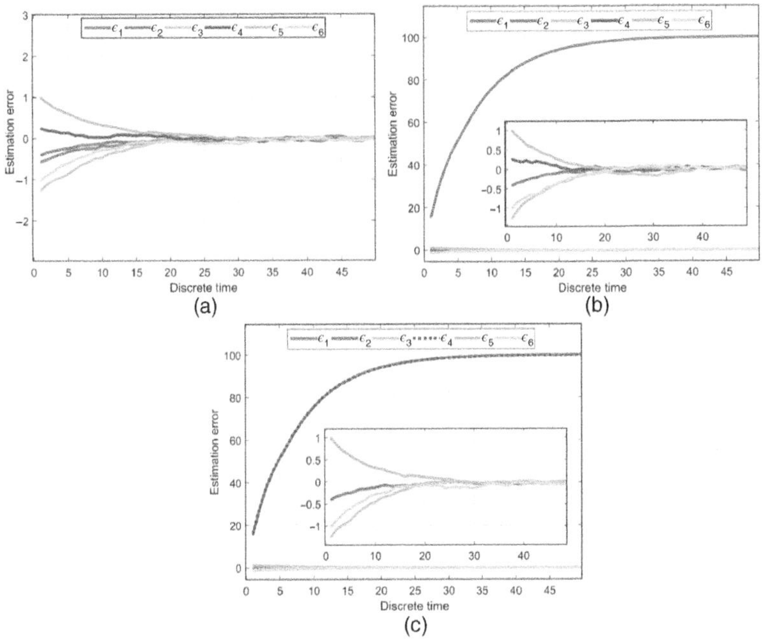

Figure 9.3 Estimator performance without and with attacks; $\gamma = 0.0001$.

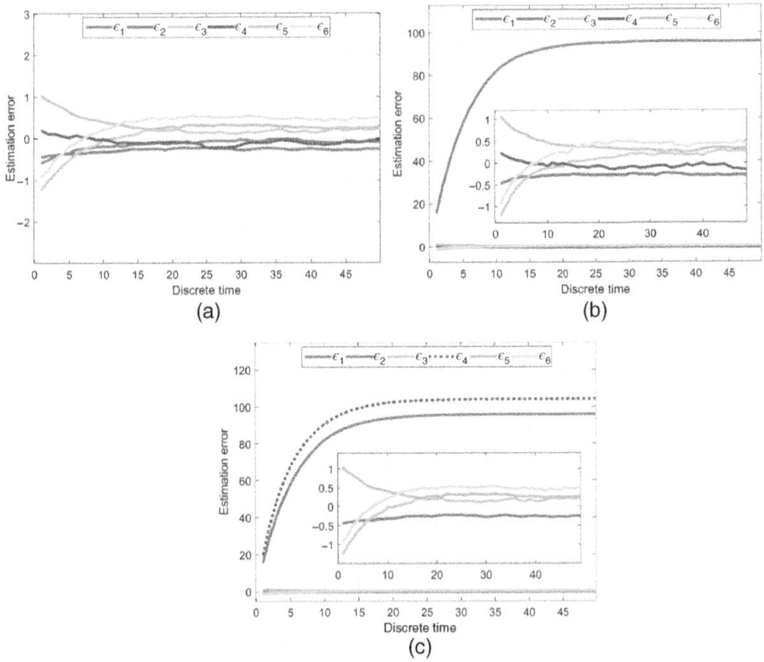

Figure 9.4 Estimator performance without and with attacks; $\gamma = 5000$.

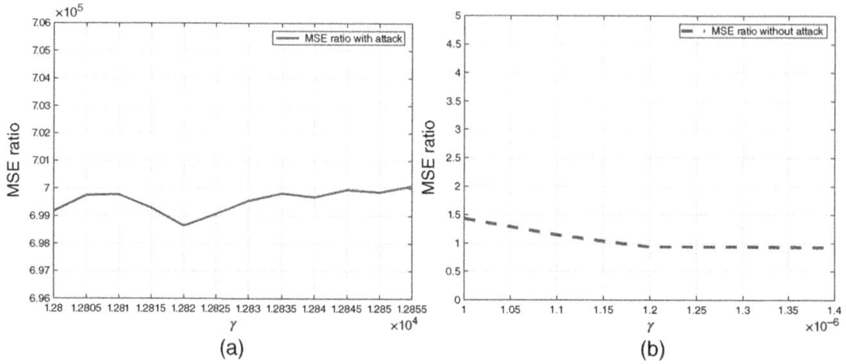

Figure 9.5 Mean squared error (MSE) ratio between the proposed secure filter and the Kalman filter with regard to parameter γ.

division are defined as the MSE ratio; the results are shown in Figure 9.5a and Figure 9.5b. When there is no attack, it can be seen that the MSE ratio is close to 1 as γ becomes larger, i.e. the proposed secure filter achieves roughly the same estimation performance as the optimal Kalman filter. For scenarios with the attack $\mathbf{a}_1(k) = \left[100, \mathbf{0}_{1\times5}\right]^{\mathrm{T}}$, it can be seen that the MSE of the proposed secure filter achieves a minimum at around $\gamma = 1.282 \times 10^4$, i.e. there is an optimal γ for the estimation performance under attack.

9.6 Conclusion

In this chapter, a secure filter was proposed to provide a secure attitude estimation framework for attitude and heading reference systems. Based on the established AHRS measuring model and the attack model, we decomposed the optimal Kalman estimate into a linear combination of local state estimates. We then proposed a convex optimization-based approach, instead of the weighted sum approach, to combine the local estimate into a more secure estimate. It was shown that the proposed secure estimator coincides with the Kalman estimator with a certain probability when there is no attack and can be stable when p elements of the model state are compromised. Finally, attacks in single and multiple states were simulated to validate the effectiveness of the proposed secure filter.

10

Conclusions

In Chapter 2, to tackle the drift issue of iterative pose estimation approaches, heading reference-assisted pose estimation was proposed to compensate for the inherent drift in visual odometry (VO) on ground vehicles, where estimation error is prone to increase when vehicles make turns and in environments with poor features. By introducing a particular orientation as a heading reference, a pose estimation framework was presented to incorporate measurements from heading reference sensors into VO. Then a graph formulation was proposed to represent the pose estimation problem under the commonly used graph optimization model.

Chapter 3 discussed scale ambiguity in monocular visual odometry (MVO). To represent measurement uncertainties, including scale ambiguity and measurement randomness, a map-assisted localization approach for mobile robots was proposed, where measurements of MVO were modeled as a group of particles that obey a uniform Gaussian or Gaussian-Gaussian distribution, depending on different measurement noise assumptions. The saliency of each particle can be obtained from the distribution to indicate the raw measurement certainty of MVO. Map-assisted shape matching is implemented as the measurement model to assign consistency to the particles generated from the distribution. Both saliency and consistency are considered in determining particle weights. Based on the statistical properties of the probability distribution, a parameter estimation scheme was proposed to narrow down the scale ambiguity of MVO while resampling particles.

In Chapter 4, a potential-field-based fusion approach was proposed for ground vehicle positioning. The potential well and potential trench were introduced to represent distinct information: GPS and VO measurements are included in potential wells, while road constraints are illustrated as potential trenches. The position estimation problem can be modeled as an optimization problem by searching for the minimum of the combined potential field. Some properties of potential fields were found and proven such that the numerical searching space could be greatly reduced. Experiments show that the proposed potential field representation is an

Multimodal Perception and Secure State Estimation for Robotic Mobility Platforms, First Edition.
Xinghua Liu, Rui Jiang, Badong Chen, and Shuzhi Sam Ge.

effective approach to multimodal information fusion and can provide accurate and robust localization results when high-precision GPS is not available.

In Chapter 5, to incorporate measurements and constraints into a unified pose estimation framework, a dynamic potential field (DPF)-based formulation was introduced to represent states, measurements, and constraints on connected Riemannian manifolds. The state equation and output equation are derived in DPF forms that imply probabilistic inference with states and measurements. Constraints are incorporated by projecting points toward the constraint subset in the state space and the measurement space, where a DPF representing the constraints is created. An information fusion scheme was designed for DPFs, which were obtained from multi-sensor measurements and constraints. It was observed that rotational drift is corrected and translational errors are reduced thanks to the fusion of stereo visual odometry (SVO), heading measurements, and road maps.

In Chapter 6, to tackle the state estimation problem for autonomous vehicles, a recursive secure dynamic estimator was designed for situations involving possible sensor attacks. We experimentally evaluated the proposed approach over multiple attack signals and showed that it outperforms conventional Kalman filtered results in the sense of offering an upper bound for error covariance in addition to estimated states.

Chapter 7 designed a recursive pose estimator inspired by the unscented Kalman filter, which can deal with the secure vehicle pose estimation problem under random deception attacks. The estimator minimizes the upper bound of the estimation error covariance, which can be solved very efficiently in real time and is suitable for recursive computation in online applications. Simulations and experiments were designed to validate the effectiveness of the proposed estimation approach.

In Chapter 8, we investigated the problem of estimating the state of a linear time-invariant Gaussian system in the presence of sparse integrity attacks. The attacker can control p out of m sensors and arbitrarily change the measurement. Moreover, the system may be unobservable by some sensors. Under mild assumptions, we prove that we can decompose the optimal Kalman estimate as a weighted sum of local state estimates. In order to find a stable state estimator against the (p, m)-sparse attack, we introduced a convex optimization-based approach to combine the local estimate into a more secure state estimate. A numerical example was provided to show that our secure estimator achieves good performance during normal operation and in attack scenarios.

In Chapter 9, a secure filter was first proposed to provide a secure attitude estimation framework for the attitude and heading reference systems. Based on the established attitude and heading reference system (AHRS) measuring model and the attack model, we decomposed the optimal Kalman estimate into a linear combination of local state estimates. A convex optimization-based approach was introduced to combine the local estimate into a more secure estimate. It was shown that the proposed secure estimator coincides with the Kalman estimator with a certain probability when there is no attack and can be stable when p elements of the model state are compromised. Attacks in single and multiple states were simulated to validate the effectiveness of the proposed secure filter.

Although some achievements were made in this book, we believe there is a long way to go to build fully autonomous vehicles, especially those with interaction, decision-making, and reasoning capabilities. Some recommendations are presented next as possible future work:

- **Lane-level pose estimation in GPS-challenged environments:** Lane-level localization is a must if pose estimation results are used as feedback for local planning and control modules on autonomous vehicles. Differential GPSs may have sub-meter accuracy but require stable GPS signals, which is not feasible in urban canyons, tunnels, and multi-layer viaducts. Moreover, the low-frequency data of GPS (around 10 Hz) would be insufficient for real-time planning and control. Recently, lane information was added for major roads in OpenStreetMap (OSM), and some methods have been proposed for automatic lane-level map generation [Guo et al., 2014]. How to minimize lateral errors in GPS-challenged environments with the help of lane-level maps is a problem worthy of study. In Chapters 3, 4, and 5, static road maps are taken as constraints. Invariant road constraints may cause difficulties for road switching at intersections. Thus, a lane-level road transition model that takes the current lane segment as a constraint would be helpful in achieving lane-level pose estimation.
- **Robust pose estimation in extreme weather and conditions:** With the development and progress of hardware, data, and algorithms, vehicles can now achieve sub-meter-level localization accuracy in ideal experimental environments. However, most of the current datasets and results are acquired in controllable weather and environments, such as daytime or sunny or cloudy days. In practice, the vehicle must be able to operate in all weather conditions with the same level of performance, especially with regard to safety. The difference between real scenarios and testing data may be misleading, and people may not realize the huge gap between theory and practice. The performance of some sensors is significantly influenced by weather and the environment. For example, cameras have poor visibility in heavy rain and snow, such that the VO and other vision-based detection algorithms may completely fail; and active sensors such as LIDAR may interfere with each other when many devices are working in the same region. These issues must be studied and solved before an L5 autonomous vehicle can be produced. In the future, it would not be surprising to see more research work and papers aiming at engineering corner cases.

- **IoV-supported joint pose estimation:** The Internet of Vehicles (IoV) is another focus area as it has benefited from 5G and telecommunication infrastructures. With IoV, in addition to self-localization, which only utilizes information from on-board sensors, more information sources could be used for robust and accurate pose estimation. For example, geographical information is inherently embedded in fixed infrastructures including lamp posts and traffic signs; and relative positions could be obtained from vehicle-to-vehicle communications. With the help of IoV, is it highly likely that future vehicles will perform better under sensor failures and extreme environments. However, it should be noted that the risk of attack increases when vehicles are connected to the IoV, when there is no effective mechanism to identify and prevent cyber attacks. Thus, closer incorporation of secure state estimation and autonomous vehicles is essential in future research and development.

References

[Agrawal and Konolige, 2006] Agrawal, M. and Konolige, K. (2006). Real-time localization in outdoor environments using stereo vision and inexpensive GPS. In *Pattern Recognition, 2006. ICPR 2006. 18th International Conference on*, volume 3, pages 1063–1068. IEEE.

[Albertini and D'Alessandro, 1996] Albertini, F. and D'Alessandro, D. (1996). Remarks on the observability of nonlinear discrete time systems. In *System Modelling and Optimization*, pages 155–162. Springer.

[Alonso et al., 2012] Alonso, I. P., Llorca, D. F., Gavilán, M., Pardo, S. Á., García-Garrido, M. Á., Vlacic, L., and Sotelo, M. Á. (2012). Accurate global localization using visual odometry and digital maps on urban environments. *Intelligent Transportation Systems, IEEE Transactions on*, 13(4):1535–1545.

[Amin et al., 2013a] Amin, S., Litrico, X., Sastry, S., and Bayen, A. M. (2013a). Cyber security of water SCADA systems-part I: Analysis and experimentation of stealthy deception attacks. *IEEE Transactions on Control Systems Technology*, 21(5):1963–1970.

[Amin et al., 2013b] Amin, S., Schwartz, G. A., and Sastry, S. S. (2013b). Security of interdependent and identical networked control systems. *Automatica*, 49(1):186–192.

[Barfoot and Furgale, 2014] Barfoot, T. D. and Furgale, P. T. (2014). Associating uncertainty with three-dimensional poses for use in estimation problems. *IEEE Transactions on Robotics*, 30(3):679–693.

[Barrow et al., 1977] Barrow, H. G., Tenenbaum, J. M., Bolles, R. C., and Wolf, H. C. (1977). Parametric correspondence and chamfer matching: Two new techniques for image matching. In *Proceedings of the 5th International Joint Conference on Artificial Intelligence - Volume 2*, IJCAI'77, pages 659–663, San Francisco, CA, USA. Morgan Kaufmann Publishers Inc.

[Barshan and Durrant-Whyte, 1995] Barshan, B. and Durrant-Whyte, H. F. (1995). Inertial navigation systems for mobile robots. *Robotics and Automation, IEEE Transactions on*, 11(3):328–342.

Multimodal Perception and Secure State Estimation for Robotic Mobility Platforms, First Edition.
Xinghua Liu, Rui Jiang, Badong Chen, and Shuzhi Sam Ge.

[BBC, 2016] BBC (2016). Hackers caused power cut in western ukraine. http://www.bbc.com/news/technology-35297464.

[Bell et al., 2009] Bell, B. M., Burke, J. V., and Pillonetto, G. (2009). An inequality constrained nonlinear Kalman-Bucy smoother by interior point likelihood maximization. *Automatica*, 45(1):25–33.

[Bezzo et al., 2014] Bezzo, N., Weimer, J., Pajic, M., Sokolsky, O., Pappas, G. J., and Lee, I. (2014). Attack resilient state estimation for autonomous robotic systems. In *Intelligent Robots and Systems (IROS 2014), 2014 IEEE/RSJ International Conference on*, pages 3692–3698. IEEE.

[Bi and Zhang, 2014] Bi, S. and Zhang, Y. J. (2014). Graphical methods for defense against false-data injection attacks on power system state estimation. *IEEE Transactions on Smart Grid*, 5(3):1216–1227.

[Biswas et al., 2006] Biswas, S., Tatchikou, R., and Dion, F. (2006). Vehicle-to-vehicle wireless communication protocols for enhancing highway traffic safety. *IEEE Communications Magazine*, 44(1):74–82.

[Blanco, 2010] Blanco, J.-L. (2010). A tutorial on SE(3) transformation parameterizations and on-manifold optimization. *University of Malaga, Tech. Rep*, 3.

[Blanco-Claraco et al., 2014] Blanco-Claraco, J.-L., Moreno-Dueñas, F.-Á., and González-Jiménez, J. (2014). The málaga urban dataset: High-rate stereo and lidar in a realistic urban scenario. *The International Journal of Robotics Research*, 33(2):207–214.

[Bloesch et al., 2017] Bloesch, M., Burri, M., Omari, S., Hutter, M., and Siegwart, R. (2017). Iterated extended Kalman filter based visual-inertial odometry using direct photometric feedback. *The International Journal of Robotics Research*, 36(10):1053–1072.

[Bloesch et al., 2015] Bloesch, M., Omari, S., Hutter, M., and Siegwart, R. (2015). Robust visual inertial odometry using a direct ekf-based approach. In *Intelligent Robots and Systems (IROS), 2015 IEEE/RSJ International Conference on*, pages 298–304.

[Boada et al., 2017] Boada, B. L., Garcia-Pozuelo, D., Boada, M. J. L., and Diaz, V. (2017). A constrained dual kalman filter based on pdf truncation for estimation of vehicle parameters and road bank angle: Analysis and experimental validation. *IEEE Transactions on Intelligent Transportation Systems*, 18(4):1006–1016.

[Borenstein and Feng, 1996] Borenstein, J. and Feng, L. (1996). Measurement and correction of systematic odometry errors in mobile robots. *Robotics and Automation, IEEE Transactions on*, 12(6):869–880.

[Borges and Aldon, 2002] Borges, G. A. and Aldon, M.-J. (2002). Optimal mobile robot pose estimation using geometrical maps. *IEEE Transactions on Robotics and Automation*, 18(1):87–94.

[Boyd and Vandenberghe, 2004] Boyd, S. and Vandenberghe, L. (2004). *Convex optimization*. Cambridge university press.

[Breu et al., 1995] Breu, H., Gil, J., Kirkpatrick, D., and Werman, M. (1995). Linear time Euclidean distance transform algorithms. *IEEE Transactions on Pattern Analysis and Machine Intelligence*, 17(5):529–533.

[Brubaker et al., 2013a] Brubaker, M. A., Geiger, A., and Urtasun, R. (2013a). Lost! leveraging the crowd for probabilistic visual self-localization. In *Proceedings of the IEEE Conference on Computer Vision and Pattern Recognition*, pages 3057–3064.

[Brubaker et al., 2013b] Brubaker, M. A., Geiger, A., and Urtasun, R. (2013b). Lost! leveraging the crowd for probabilistic visual self-localization. In *IEEE Conf. on Computer Vision and Pattern Recognition (CVPR 2013)*, pages 3057–3064, Portland, OR. IEEE.

[Cárdenas et al., 2008] Cárdenas, A. A., Amin, S., and Sastry, S. (2008). Research challenges for the security of control systems. In *Proc. Conf. Hot Topics in Security*, pages 1–6, Berkeley, CA, USA. USENIX Association.

[Cardenas et al., 2008] Cardenas, A. A., Amin, S., and Sastry, S. (2008). Secure control: Towards survivable cyber-physical systems. In *Distributed Computing Systems Workshops, 2008 IEEE 28th International Conference on*, pages 495–500.

[Carlevaris-Bianco and Eustice, 2013] Carlevaris-Bianco, N. and Eustice, R. M. (2013). Generic factor-based node marginalization and edge sparsification for pose-graph SLAM. In *Robotics and Automation (ICRA), 2013 IEEE International Conference on*, pages 5748–5755. IEEE.

[Carlevaris-Bianco et al., 2014] Carlevaris-Bianco, N., Kaess, M., and Eustice, R. M. (2014). Generic node removal for factor-graph SLAM. *IEEE Transactions on Robotics*, 30(6):1371–1385.

[Castellanos and Tardos, 2012] Castellanos, J. A. and Tardos, J. D. (2012). *Mobile robot localization and map building: A multisensor fusion approach*. Springer Science & Business Media.

[Chen, 2010] Chen, T. M. (2010). Stuxnet, the real start of cyber warfare? [editor's note]. *IEEE Network*, 24(6):2–3.

[Chen and Zheng, 2007a] Chen, W.-H. and Zheng, W. X. (2007a). Delay-dependent robust stabilization for uncertain neutral systems with distributed delays. *Automatica*, 43(1):95–104.

[Chen and Zheng, 2007b] Chen, W.-H. and Zheng, W.-X. (2007b). Delay-dependent robust stabilization for uncertain neutral systems with distributed delays. *Automatica*, 43(1):95–104.

[Chen et al., 2018] Chen, Y., Peng, H., and Grizzle, J. (2018). Obstacle avoidance for low-speed autonomous vehicles with barrier function. *IEEE Transactions on Control Systems Technology*, 26(1):194–206.

[Chen et al., 2016] Chen, Z., Zhu, Q., and Soh, Y. C. (2016). Smartphone inertial sensor-based indoor localization and tracking with ibeacon corrections. *IEEE Transactions on Industrial Informatics*, 12(4):1540–1549.

[Cheng et al., 2005] Cheng, Y., Maimone, M., and Matthies, L. (2005). Visual odometry on the Mars exploration rovers. In *Systems, Man and Cybernetics, 2005 IEEE International Conference on*, volume 1, pages 903–910. IEEE.

[Chirikjian, 2009] Chirikjian, G. S. (2009). *Stochastic Models, Information Theory, and Lie Groups, Volume 1: Classical Results and Geometric Methods*, volume 1. Springer Science & Business Media.

[Chirikjian, 2011] Chirikjian, G. S. (2011). *Stochastic Models, Information Theory, and Lie Groups, Volume 2: Analytic Methods and Modern Applications*, volume 2. Springer Science & Business Media.

[Choset, 2005] Choset, H. M. (2005). *Principles of robot motion: theory, algorithms, and implementation*. MIT press.

[Costanzi et al., 2016] Costanzi, R., Fanelli, F., Monni, N., Ridolfi, A., and Allotta, B. (2016). An attitude estimation algorithm for mobile robots under unknown magnetic disturbances. *IEEE/ASME Transactions on Mechatronics*, 21(4):1900–1911.

[Crassidis, 2006] Crassidis, J. L. (2006). Sigma-point kalman filtering for integrated gps and inertial navigation. *Aerospace and Electronic Systems, IEEE Transactions on*, 42(2):750–756.

[Cui and Ge, 2003] Cui, Y. and Ge, S. S. (2003). Autonomous vehicle positioning with gps in urban canyon environments. *IEEE transactions on robotics and automation*, 19(1):15–25.

[Davison, 2003] Davison, A. J. (2003). Real-time simultaneous localisation and mapping with a single camera. In *Computer Vision, 2003. Proceedings. Ninth IEEE International Conference on*, pages 1403–1410. IEEE.

[Ding et al., 2018a] Ding, D., Wang, Z., Han, Q., and Wei, G. (2018a). Security control for discrete-time stochastic nonlinear systems subject to deception attacks. *IEEE Transactions on Systems, Man, and Cybernetics: Systems*, 48(5):779–789.

[Ding et al., 2016] Ding, D., Wang, Z., Han, Q.-L., and Wei, G. (2016). Security control for discrete-time stochastic nonlinear systems subject to deception attacks. *IEEE Transactions on Systems, Man, and Cybernetics: Systems*.

[Ding et al., 2018b] Ding, D., Wang, Z., Han, Q.-L., and Wei, G. (2018b). Security control for discrete-time stochastic nonlinear systems subject to deception attacks. *IEEE Transactions on Systems, Man, and Cybernetics: Systems*, 48(5):779–789.

[Ding et al., 2020] Ding, X., Wang, Z., Zhang, L., and Wang, C. (2020). Longitudinal vehicle speed estimation for four-wheel-independently-actuated electric vehicles based on multi-sensor fusion. *IEEE Transactions on Vehicular Technology*, 69(11):12797–12806.

[El Najjar and Bonnifait, 2005] El Najjar, M. E. and Bonnifait, P. (2005). A road-matching method for precise vehicle localization using belief theory and kalman filtering. *Autonomous Robots*, 19(2):173–191.

[Estrada et al., 2005] Estrada, C., Neira, J., and Tardós, J. D. (2005). Hierarchical SLAM: Real-time accurate mapping of large environments. *IEEE Transactions on Robotics*, 21(4):588–596.

[Fagnant and Kockelman, 2015] Fagnant, D. J. and Kockelman, K. (2015). Preparing a nation for autonomous vehicles: opportunities, barriers and policy recommendations. *Transportation Research Part A: Policy and Practice*, 77:167–181.

[Falquez et al., 2016] Falquez, J. M., Kasper, M., and Sibley, G. (2016). Inertial aided dense & semi-dense methods for robust direct visual odometry. In *Intelligent Robots and Systems (IROS), 2016 IEEE/RSJ International Conference on*, pages 3601–3607. IEEE.

[Farahmand et al., 2011] Farahmand, S., Giannakis, G. B., and Angelosante, D. (2011). Doubly robust smoothing of dynamical processes via outlier sparsity constraints. *IEEE Transactions on Signal Processing*, 59(10):4529–4543.

[Farwell and Rohozinski, 2011] Farwell, J. P. and Rohozinski, R. (2011). Stuxnet and the future of cyber war. *Survival*, 53(1):23–40.

[Fawzi et al., 2011] Fawzi, H., Tabuada, P., and Diggavi, S. (2011). Secure state-estimation for dynamical systems under active adversaries. In *2011 49th Annual Allerton Conference on Communication, Control, and Computing (Allerton)*, Monticello, IL, pages 337–344.

[Fawzi et al., 2014a] Fawzi, H., Tabuada, P., and Diggavi, S. (2014a). Secure estimation and control for cyber-physical systems under adversarial attacks. *IEEE Transactions on Automatic Control*, 59(6):1454–1467.

[Fawzi et al., 2014b] Fawzi, H., Tabuada, P., and Diggavi, S. (2014b). Secure estimation and control for cyber-physical systems under adversarial attacks. *IEEE Trans. Autom. Control*, 59(6):1454–1467.

[Fidler, 2011] Fidler, D. P. (2011). Was stuxnet an act of war? decoding a cyberattack. *IEEE Security & Privacy*, 9(4):56–59.

[Floros et al., 2013] Floros, G., van der Zander, B., and Leibe, B. (2013). Openstreetslam: Global vehicle localization using openstreetmaps. In *Robotics and Automation (ICRA), 2013 IEEE International Conference on*, pages 1054–1059. IEEE.

[Forster et al., 2017] Forster, C., Carlone, L., Dellaert, F., and Scaramuzza, D. (2017). On-manifold preintegration for real-time visual-inertial odometry. *IEEE Transactions on Robotics*, 33(1):1–21.

[Ge and Fua, 2005] Ge, S. S. and Fua, C.-H. (2005). Queues and artificial potential trenches for multirobot formations. *Robotics, IEEE Transactions on*, 21(4): 646–656.

[Ge et al., 2016] Ge, S. S., Liu, X., Goh, C.-H., and Xu, L. (2016). Formation tracking control of multiagents in constrained space. *IEEE Transactions on Control Systems Technology*, 24(3):992–1003.

[Geiger et al., 2013] Geiger, A., Lenz, P., Stiller, C., and Urtasun, R. (2013). Vision meets robotics: The kitti dataset. *The International Journal of Robotics Research*, page 0278364913491297.

[Geiger et al., 2012] Geiger, A., Lenz, P., and Urtasun, R. (2012). Are we ready for autonomous driving? the kitti vision benchmark suite. In *Conference on Computer Vision and Pattern Recognition (CVPR)*.

[Geiger et al., 2011a] Geiger, A., Ziegler, J., and Stiller, C. (2011a). Stereoscan: Dense 3D reconstruction in real-time. In *Intelligent Vehicles Symposium (IV), 2011 IEEE*, pages 963–968. IEEE.

[Geiger et al., 2011b] Geiger, A., Ziegler, J., and Stiller, C. (2011b). Stereoscan: Dense 3d reconstruction in real-time. In *Intelligent Vehicles Symposium (IV)*.

[Gerla et al., 2014] Gerla, M., Lee, E.-K., Pau, G., and Lee, U. (2014). Internet of vehicles: From intelligent grid to autonomous cars and vehicular clouds. In *Internet of Things (WF-IoT), 2014 IEEE World Forum on*, pages 241–246. IEEE.

[Gerland, 1993] Gerland, H. E. (1993). Its intelligent transportation system: fleet management with gps dead reckoning, advanced displays, smartcards, etc. In *Vehicle Navigation and Information Systems Conference, 1993., Proceedings of the IEEE-IEE*, pages 606–611. IEEE.

[Ghabcheloo et al., 2006] Ghabcheloo, R., Aguiar, A. P., Pascoal, A., Silvestre, C., Kaminer, I., and Hespanha, J. (2006). Coordinated path-following control of multiple underactuated autonomous vehicles in the presence of communication failures. In *Decision and Control, 2006 45th IEEE Conference on*, pages 4345–4350. IEEE.

[Gope et al., 2017] Gope, P., Lee, J., and Quek, T. Q. S. (2017). Resilience of DoS attacks in designing anonymous user authentication protocol for wireless sensor networks. *IEEE Sensors Journal*, 17(2):498–503.

[Grant and Boyd, 2014] Grant, M. and Boyd, S. (2014). CVX: Matlab software for disciplined convex programming, version 2.1. http://cvxr.com/cvx.

[Grisetti et al., 2010] Grisetti, G., Kummerle, R., Stachniss, C., and Burgard, W. (2010). A tutorial on graph-based SLAM. *IEEE Intelligent Transportation Systems Magazine*, 2(4):31–43.

[Guo et al., 2014] Guo, C., Meguro, J.-i., Kojima, Y., and Naito, T. (2014). Automatic lane-level map generation for advanced driver assistance systems using low-cost sensors. In *Robotics and Automation (ICRA), 2014 IEEE International Conference on*, pages 3975–3982. IEEE.

[Guo et al., 2020] Guo, J., Jiang, R., He, B., Yan, T., and Ge, S. S. (2020). General learning modeling for auv position tracking. *IEEE Intelligent Systems*.

[Hajiyev et al., 2012] Hajiyev, C., Ata, M., Dinc, M., and Soken, H. E. (2012). Fault tolerant estimation of autonomous underwater vehicle dynamics via robust UKF. In *Proceedings of the 13th International Carpathian Control Conference (ICCC)*, pages 203–208.

[Hall and Llinas, 2001] Hall, D. and Llinas, J. (2001). *Multisensor data fusion*. CRC press.

[Han et al., 2015] Han, D., Mo, Y., and Xie, L. (2015). Convex optimization based state estimation against sparse integrity attacks. *arXiv preprint arXiv:1511.07218*.

[Han et al., 2016] Han, D., Mo, Y., and Xie, L. (2016). Convex optimization based state estimation against sparse integrity attacks.

[Han and Ge, 2015] Han, T. T. and Ge, S. S. (2015). Styled-velocity flocking of autonomous vehicles: A systematic design. *IEEE Transactions on Automatic Control*, 60(8):2015–2030.

[Hansen and Blanke, 2014] Hansen, S. and Blanke, M. (2014). Diagnosis of airspeed measurement faults for unmanned aerial vehicles. *IEEE Transactions on Aerospace and Electronic Systems*, 50(1):224–239.

[Hawking and Ellis, 1973] Hawking, S. W. and Ellis, G. F. R. (1973). *The large scale structure of space-time*, volume 1. Cambridge university press.

[Hermann and Krener, 1977] Hermann, R. and Krener, A. (1977). Nonlinear controllability and observability. *IEEE Transactions on Automatic Control*, 22(5):728–740.

[Hespanha et al., 2007] Hespanha, J. P., Naghshtabrizi, P., and Xu, Y. (2007). A survey of recent results in networked control systems. *Proceedings of the IEEE*, 95(1):138–162.

[Hoang et al., 2017] Hoang, V.-D., Le, M.-H., and Jo, K.-H. (2017). Motion estimation based on two corresponding points and angular deviation optimization. *IEEE Transactions on Industrial Electronics*, 64(11):8598–8606.

[Hu and Qiao, 2016] Hu, G.-Y. and Qiao, P.-L. (2016). Cloud belief rule base model for network security situation prediction. *IEEE Communications Letters*, 20(5):914–917.

[Hu and Sun, 2015] Hu, J.-S. and Sun, K.-C. (2015). A robust orientation estimation algorithm using MARG sensors. *IEEE Transactions on Instrumentation and Measurement*, 64(3):815–822.

[Huang et al., 2017] Huang, A. S., Bachrach, A., Henry, P., Krainin, M., Maturana, D., Fox, D., and Roy, N. (2017). Visual odometry and mapping for autonomous flight using an RGB-D camera. In *Robotics Research*, pages 235–252. Springer.

[Huang and Pi, 2011] Huang, P. and Pi, Y. (2011). Urban environment solutions to gps signal near-far effect. *Aerospace and Electronic Systems Magazine, IEEE*, 26(5):18–27.

[Huang and Chiang, 2008] Huang, Y.-W. and Chiang, K.-W. (2008). An intelligent and autonomous MEMS IMU/GPS integration scheme for low cost land navigation applications. *GPS Solutions*, 12(2):135–146.

[Huber and Ronchetti, 2009] Huber, P. J. and Ronchetti, E. M. (2009). *Robust Statistics*. NJ:Wiely.

[Inoue et al., 2016] Inoue, R. S., Terra, M. H., and Cerri, J. P. (2016). Extended robust kalman filter for attitude estimation. *IET Control Theory & Applications*, 10(2):162–172.

[International, 2016] International, S. (2016). Taxonomy and definitions for terms related to driving automation systems for on-road motor vehicles.

[Jagadeesh et al., 2004] Jagadeesh, G., Srikanthan, T., and Zhang, X. (2004). A map matching method for gps based real-time vehicle location. *Journal of Navigation*, 57(03):429–440.

[Janabi-Sharifi and Marey, 2010] Janabi-Sharifi, F. and Marey, M. (2010). A Kalman-filter-based method for pose estimation in visual servoing. *IEEE Transactions on Robotics*, 26(5):939–947.

[Jiang et al., 2010] Jiang, R., Klette, R., and Wang, S. (2010). Modeling of unbounded long-range drift in visual odometry. In *Image and Video Technology (PSIVT), 2010 Fourth Pacific-Rim Symposium on*, pages 121–126. IEEE.

[Jiang et al., 2019a] Jiang, R., Liu, X., Wang, H., and Ge, S. (2019a). Secure estimation for attitude and heading reference systems under sparse attacks. *IEEE Sensors Journal*, 19(2):641–649.

[Jiang et al., 2019b] Jiang, R., Liu, X., Wang, H., and Ge, S. S. (2019b). Secure estimation for attitude and heading reference systems under sparse attacks. *IEEE Sensors Journal*, 19(2):641–649.

[Jiang et al., 2018] Jiang, R., Yang, S., Ge, S. S., Liu, X., Wang, H., and Lee, T. H. (2018). GPS/odometry/map fusion for vehicle positioning using potential function. *Autonomous Robots*, 42(1):99–110.

[Jiang et al., 2017] Jiang, R., Yang, S., Ge, S. S., Wang, H., and Lee, T. H. (2017). Geometric map-assisted localization for mobile robots based on uniform-gaussian distribution. *IEEE Robotics and Automation Letters*, 2(2):789–795.

[Jiang et al., 2019c] Jiang, R., Zhou, H., Wang, H., and Ge, S. S. (2019c). Road-constrained geometric pose estimation for ground vehicles. *IEEE Transactions on Automation Science and Engineering*, 17(2):748–760.

[Jiang et al., 2020] Jiang, R., Zhou, H., Wang, H., and Ge, S. S. (2020). Road-constrained geometric pose estimation for ground vehicles. *IEEE Transactions on Automation Science and Engineering*, 17(2):748–760.

[Jiang et al., 2009] Jiang, T., Jurie, F., and Schmid, C. (2009). Learning shape prior models for object matching. In *Computer Vision and Pattern Recognition, 2009. CVPR 2009. IEEE Conference on*, pages 848–855. IEEE.

[Jo et al., 2012] Jo, K., Chu, K., and Sunwoo, M. (2012). Interacting multiple model filter-based sensor fusion of gps with in-vehicle sensors for real-time vehicle positioning. *Intelligent Transportation Systems, IEEE Transactions on*, 13(1):329–343.

[Jo et al., 2016] Jo, K., Lee, M., and Sunwoo, M. (2016). Road slope aided vehicle position estimation system based on sensor fusion of GPS and automotive onboard sensors. *IEEE Transactions on Intelligent Transportation Systems*, 17(1):250–263.

[Joo et al., 2009] Joo, H., Jeong, Y., Duchenne, O., Ko, S.-Y., and Kweon, I.-S. (2009). Graph-based robust shape matching for robotic application. In *Robotics and Automation, 2009. ICRA'09. IEEE International Conference on*, pages 1207–1213. IEEE.

[Ju et al., 2020] Ju, Z., Zhang, H., and Tan, Y. (2020). Deception attack detection and estimation for a local vehicle in vehicle platooning based on a modified ufir estimator. *IEEE Internet of Things Journal*, 7(5):3693–3705.

[Kaess et al., 2007] Kaess, M., Ranganathan, A., and Dellaert, F. (2007). iSAM: Fast incremental smoothing and mapping with efficient data association. In *Robotics and Automation, 2007 IEEE International Conference on*, pages 1670–1677. IEEE.

[Kaplan and Hegarty, 2005] Kaplan, E. and Hegarty, C. (2005). *Understanding GPS: principles and applications*. Artech house.

[Karcher, 1977] Karcher, H. (1977). Riemannian center of mass and mollifier smoothing. *Communications on Pure and Applied Mathematics*, 30(5):509–541.

[Kassam et al., 1985] Kassam, S., Poor, H. V., et al. (1985). Robust techniques for signal processing: A survey. *Proc. IEEE*, 73(3):433–481.

[Kavousi-Fard et al., 2016] Kavousi-Fard, A., Niknam, T., and Fotuhi-Firuzabad, M. (2016). A novel stochastic framework based on cloud theory and-modified bat algorithm to solve the distribution feeder reconfiguration. *IEEE Transactions on Smart Grid*, 7(2):740–750.

[Kelly, 2013] Kelly, A. (2013). *Mobile Robotics: Mathematics, Models, and Methods*. Cambridge University Press.

[Kendall, 1990] Kendall, W. S. (1990). Probability, convexity, and harmonic maps with small image I: uniqueness and fine existence. *Proceedings of the London Mathematical Society*, 3(2):371–406.

[Khaitan and McCalley, 2015] Khaitan, S. K. and McCalley, J. D. (2015). Design techniques and applications of cyberphysical systems: A survey. *IEEE Systems Journal*, 9(2):350–365.

[Khatib, 1986] Khatib, O. (1986). Real-time obstacle avoidance for manipulators and mobile robots. *The international journal of robotics research*, 5(1):90–98.

[Kitt et al., 2011] Kitt, B. M., Rehder, J., Chambers, A. D., Schonbein, M., Lategahn, H., and Singh, S. (2011). Monocular visual odometry using a planar road model to solve scale ambiguity.

[Kong, 2011] Kong, S.-H. (2011). Statistical analysis of urban gps multipaths and pseudo-range measurement errors. *Aerospace and Electronic Systems, IEEE Transactions on*, 47(2):1101–1113.

[Konolige and Garage, 2010] Konolige, K. and Garage, W. (2010). Sparse sparse bundle adjustment. In *BMVC*, volume 10, pages 102–1. Citeseer.

[Kothari et al., 2017] Kothari, N., Gupta, M., Vachhani, L., and Arya, H. (2017). Pose estimation for an autonomous vehicle using monocular vision. In *2017 Indian Control Conference (ICC)*, pages 424–431.

[Koval et al., 2015] Koval, M. C., Pollard, N. S., and Srinivasa, S. S. (2015). Pose estimation for planar contact manipulation with manifold particle filters. *The International Journal of Robotics Research*, 34(7):922–945.

[Kümmerle et al., 2011] Kümmerle, R., Grisetti, G., Strasdat, H., Konolige, K., and Burgard, W. (2011). g^2o: A general framework for graph optimization. In *Robotics and Automation (ICRA), 2011 IEEE International Conference on*, pages 3607–3613. IEEE.

[Lam et al., 2016] Lam, A. Y. S., Leung, Y. W., and Chu, X. (2016). Autonomous-vehicle public transportation system: Scheduling and admission control. *IEEE Transactions on Intelligent Transportation Systems*, 17(5):1210–1226.

[Lee et al., 2015] Lee, C., Shim, H., and Eun, Y. (2015). Secure and robust state estimation under sensor attacks, measurement noises, and process disturbances: Observer-based combinatorial approach. In *Control Conference (ECC), 2015 European*, pages 1872–1877. IEEE.

[Leutenegger et al., 2015] Leutenegger, S., Lynen, S., Bosse, M., Siegwart, R., and Furgale, P. (2015). Keyframe-based visual–inertial odometry using nonlinear optimization. *The International Journal of Robotics Research*, 34(3):314–334.

[Levinson et al., 2007] Levinson, J., Montemerlo, M., and Thrun, S. (2007). Map-based precision vehicle localization in urban environments. In *Robotics: Science and Systems*, volume 4, page 1. Citeseer.

[Li and Du, 2007] Li, D. and Du, Y. (2007). *Artificial intelligence with uncertainty*. CRC press.

[Li et al., 2009] Li, D., Liu, C., and Gan, W. (2009). A new cognitive model: cloud model. *International Journal of Intelligent Systems*, 24(3):357–375.

[Li and Mourikis, 2013] Li, M. and Mourikis, A. I. (2013). High-precision, consistent EKF-based visual-inertial odometry. *The International Journal of Robotics Research*, 32(6):690–711.

[Li et al., 2014] Li, Q., Chen, L., Zhu, Q., Li, M., Zhang, Q., and Ge, S. S. (2014). Intersection detection and recognition for autonomous urban driving using a virtual cylindrical scanner. *IET Intelligent Transport Systems*, 8(3):244–254.

[Li et al., 2015] Li, Y., Shi, L., Cheng, P., Chen, J., and Quevedo, D. E. (2015). Jamming attacks on remote state estimation in cyber-physical systems: A game-theoretic approach. *IEEE Transactions on Automatic Control*, 60(10):2831–2836.

[Lim et al., 2015] Lim, H., Sinha, S. N., Cohen, M. F., Uyttendaele, M., and Kim, H. J. (2015). Real-time monocular image-based 6-dof localization. *The International Journal of Robotics Research*, 34(4-5):476–492.

[Liu et al., 2020] Liu, J., Wang, Z., Zhang, L., and Walker, P. (2020). Sideslip angle estimation of ground vehicles: a comparative study. *IET Control Theory & Applications*, 14(20):3490–3505.

[Liu et al., 2018a] Liu, J., Xia, J., Tian, E., and Fei, S. (2018a). Hybrid-driven-based H_∞ filter design for neural networks subject to deception attacks. *Applied Mathematics and Computation*, 320:158–174.

[Liu et al., 2018b] Liu, L., Li, H., Dai, Y., and Pan, Q. (2018b). Robust and efficient relative pose with a multi-camera system for autonomous driving in highly dynamic environments. *IEEE Transactions on Intelligent Transportation Systems*, 19(8):2432–2444.

[Liu and Shi, 2013] Liu, M. and Shi, P. (2013). Sensor fault estimation and tolerant control for Itô stochastic systems with a descriptor sliding mode approach. *Automatica*, 49(5):1242–1250.

[Liu et al., 2019a] Liu, Q., Mo, Y., Mo, X., Lv, C., Mihankhah, E., and Wang, D. (2019a). Secure pose estimation for autonomous vehicles under cyber attacks. In *2019 IEEE Intelligent Vehicles Symposium (IV), Paris, France*, pages 1583–1588.

[Liu et al., 2021] Liu, X., Bai, D., Lv, Y., Jiang, R., and Ge, S. (2021). Ukf-based vehicle pose estimation under randomly occurring deception attacks. *Security and Communication Networks*, 2021, Article ID 5572186:1–12.

[Liu et al., 2019b] Liu, X., Jiang, R., Wang, H., and Ge, S. (2019b). Filter-based secure dynamic pose estimation for autonomous vehicles. *IEEE Sensors Journal*, 19(15):6298–6308.

[Liu et al., 2019c] Liu, X., Jiang, R., Wang, H., and Ge, S. S. (2019c). Filter-based secure dynamic pose estimation for autonomous vehicles. *IEEE Sensors Journal*, 19(15):6298–6308.

[Liu et al., 2017] Liu, X., Mo, Y., and Garone, E. (2017). Secure dynamic state estimation by decomposing kalman filter. *IFAC-PapersOnLine*, 50(1):7351–7356.

[Liu et al., 2009] Liu, Y., Reiter, M., and Ning, P. (2009). False data injection attacks against state estimation in electric power grids. In *Proc. ACM Conf. Computer and Commun. Security*.

[Liu et al., 2016] Liu, Y., Xiong, R., Wang, Y., Huang, H., Xie, X., Liu, X., and Zhang, G. (2016). Stereo visual-inertial odometry with multiple kalman filters ensemble. *IEEE Transactions on Industrial Electronics*, 63(10):6205–6216.

[Lobo and Dias, 2003] Lobo, J. and Dias, J. (2003). Vision and inertial sensor cooperation using gravity as a vertical reference. *IEEE Transactions on Pattern Analysis and Machine Intelligence*, 25(12):1597–1608.

[Loureiro et al., 2014] Loureiro, R., Benmoussa, S., Touati, Y., Merzouki, R., and Bouamama, B. O. (2014). Integration of fault diagnosis and fault-tolerant control for health monitoring of a class of MIMO intelligent autonomous vehicles. *IEEE Transactions on Vehicular Technology*, 63(1):30–39.

[Lu et al., 2017] Lu, Y., Huang, J., Chen, Y.-T., and Heisele, B. (2017). Monocular localization in urban environments using road markings. In *Intelligent Vehicles Symposium (IV), 2017 IEEE*, pages 468–474. IEEE.

[Lupton and Sukkarieh, 2012] Lupton, T. and Sukkarieh, S. (2012). Visual-inertial-aided navigation for high-dynamic motion in built environments without initial conditions. *IEEE Transactions on Robotics*, 28(1):61–76.

[Ma et al., 2017a] Ma, L., Wang, Z., Han, Q.-L., and Lam, H.-K. (2017a). Variance-constrained distributed filtering for time-varying systems with multiplicative noises and deception attacks over sensor networks. *IEEE Sensors Journal*, 17(7):2279–2288.

[Ma et al., 2017b] Ma, L., Wang, Z., Lam, H., and Kyriakoulis, N. (2017b). Distributed event-based set-membership filtering for a class of nonlinear systems with sensor saturations over sensor networks. *IEEE Transactions on Cybernetics*, 47(11):3772–3783.

[Ma and Fu, 2011] Ma, Y. and Fu, Y. (2011). *Manifold learning theory and applications*. CRC press.

[Maimone et al., 2007] Maimone, M., Cheng, Y., and Matthies, L. (2007). Two years of visual odometry on the mars exploration rovers. *Journal of Field Robotics*, 24(3):169–186.

[Markley and Mortari, 2000] Markley, F. L. and Mortari, D. (2000). Quaternion attitude estimation using vector observations. *Journal of the Astronautical Sciences*, 48(2):359–380.

[Maronna et al., 2006] Maronna, R. A., Martin, D. R., and Yohai, V. J. (2006). *Robust Statistics: Theory and Methods*. NJ: Wiley.

[Mattingley and Boyd, 2010] Mattingley, J. and Boyd, S. (2010). Real-time convex optimization in signal processing. *IEEE Signal Processing Magazine*, 27(3):50–61.

[Meng and Li, 2018] Meng, C. and Li, W. (2018). Recursive filtering for complex networks against random deception attacks. In *2018 IEEE International Conference on Big Data and Smart Computing (BigComp), Shanghai*, pages 565–568.

[Merriaux et al., 2015] Merriaux, P., Dupuis, Y., Vasseur, P., and Savatier, X. (2015). Fast and robust vehicle positioning on graph-based representation of drivable maps. In *2015 IEEE International Conference on Robotics and Automation (ICRA)*, pages 2787–2793. IEEE.

[Milford et al., 2014] Milford, M., Vig, E., Scheirer, W., and Cox, D. (2014). Vision-based simultaneous localization and mapping in changing outdoor environments. *Journal of Field Robotics*, 31(5):780–802.

[Miranda and Edelmayer, 2012] Miranda, M. and Edelmayer, A. (2012). Fault tolerant adaptive estimation of nonlinear processes using redundant measurements. In *20th Mediterranean Conference on Control Automation (MED)*, pages 703–709.

[Mishra et al., 2017] Mishra, S., Shoukry, Y., Karamchandani, N., Diggavi, S. N., and Tabuada, P. (2017). Secure state estimation against sensor attacks in the presence of noise. *IEEE Transactions on Control of Network Systems*, 4(1):49–59.

[Mo and Garone, 2016] Mo, Y. and Garone, E. (Dec 2016). Secure dynamic state estimation via local estimators. In *Decision and Control (CDC), 2016 IEEE 55th Conference on*, pages 5073–5078.

[Mo and Sinopoli, 2010] Mo, Y. and Sinopoli, B. (2010). False data injection attacks in cyber physical systems. In *First Workshop on Secure Control Systems*.

[Mo and Sinopoli, 2015a] Mo, Y. and Sinopoli, B. (2015a). On the performance degradation of cyber-physical systems under stealthy integrity attacks. *IEEE Transactions on Automatic Control*, PP(99):1–1.

[Mo and Sinopoli, 2015b] Mo, Y. and Sinopoli, B. (2015b). Secure estimation in the presence of integrity attacks. *IEEE Transactions on Automatic Control*, 60(4):1145–1151.

[Mourikis and Roumeliotis, 2007] Mourikis, A. I. and Roumeliotis, S. I. (2007). A multi-state constraint Kalman filter for vision-aided inertial navigation. In *Robotics and automation, 2007 IEEE international conference on*, pages 3565–3572. IEEE.

[Mur-Artal et al., 2015a] Mur-Artal, R., Montiel, J., and Tardós, J. D. (2015a). ORB-SLAM: a versatile and accurate monocular SLAM system. *IEEE Transactions on Robotics*, 31(5):1147–1163.

[Mur-Artal et al., 2015b] Mur-Artal, R., Montiel, J. M. M., and Tardós, J. D. (2015b). Orb-slam: A versatile and accurate monocular slam system. *IEEE Transactions on Robotics*, 31(5):1147–1163.

[Nistér et al., 2006] Nistér, D., Naroditsky, O., and Bergen, J. (2006). Visual odometry for ground vehicle applications. *Journal of Field Robotics*, 23(1):3–20.

[Noureldin et al., 2009] Noureldin, A., Karamat, T. B., Eberts, M. D., and El-Shafie, A. (2009). Performance enhancement of mems-based ins/gps integration for low-cost navigation applications. *Vehicular Technology, IEEE Transactions on*, 58(3):1077–1096.

[Nummiaro et al., 2003] Nummiaro, K., Koller-Meier, E., and Van Gool, L. (2003). An adaptive color-based particle filter. *Image and Vision Computing*, 21(1):99–110.

[Nützi et al., 2011] Nützi, G., Weiss, S., Scaramuzza, D., and Siegwart, R. (2011). Fusion of IMU and vision for absolute scale estimation in monocular SLAM. *Journal of Intelligent & Robotic Systems*, 61(1-4):287–299.

[Pajic et al., 2015] Pajic, M., Tabuada, P., Lee, I., and Pappas, G. J. (2015). Attack-resilient state estimation in the presence of noise. In *2015 54th IEEE Conference on Decision and Control (CDC)*, pages 5827–5832.

[Pajic et al., 2014a] Pajic, M., Weimer, J., Bezzo, N., Tabuada, P., Sokolsky, O., Lee, I., and Pappas, G. J. (2014a). Robustness of attack-resilient state estimators. In *Proc. ACM/IEEE Int. Conf. Cyber-Physical Systems*, pages 163–174.

[Pajic et al., 2014b] Pajic, M., Weimer, J., Bezzo, N., Tabuada, P., Sokolsky, O., Lee, I., and Pappas, G. J. (2014b). Robustness of attack-resilient state estimators. In *ICCPS'14: ACM/IEEE 5th International Conference on Cyber-Physical Systems (with CPS Week 2014)*, pages 163–174. IEEE.

[Parent, 2007] Parent, M. (2007). Advanced urban transport: Automation is on the way. *IEEE Intelligent Systems*, 22(2).

[Parisotto et al., 2018] Parisotto, E., Chaplot, D. S., Zhang, J., and Salakhutdinov, R. (2018). Global pose estimation with an attention-based recurrent network. *arXiv preprint arXiv:1802.06857*.

[Parkinson et al., 2017] Parkinson, S., Ward, P., Wilson, K., and Miller, J. (2017). Cyber threats facing autonomous and connected vehicles: future challenges. *IEEE Transactions on Intelligent Transportation Systems*, 18(11):2898–2915.

[Pasqualetti et al., 2010] Pasqualetti, F., Bicchi, A., and Bullo, F. (2010). Consensus computation in unreliable networks: a system theoretic approach. *IEEE Trans. Autom. Control*, 57(1):90–104.

[Pasqualetti et al., 2013] Pasqualetti, F., Dörfler, F., and Bullo, F. (2013). Attack detection and identification in cyber-physical systems. *IEEE Transactions on Automatic Control*, 58(11):2715–2729.

[Pasqualetti et al., 2015] Pasqualetti, F., Dorfler, F., and Bullo, F. (2015). Control-theoretic methods for cyberphysical security: Geometric principles for optimal cross-layer resilient control systems. *IEEE Control Systems*, 35(1):110–127.

[Pennec, 2004] Pennec, X. (2004). *Probabilities and statistics on riemannian manifolds: A geometric approach*. PhD thesis, INRIA.

[Philip et al., 2018] Philip, B. V., Alpcan, T., Jin, J., and Palaniswami, M. (2018). Distributed real-time iot for autonomous vehicles. *IEEE Transactions on Industrial Informatics*, 15(2):1131–1140.

[Piniés et al., 2007] Piniés, P., Lupton, T., Sukkarieh, S., and Tardós, J. D. (2007). Inertial aiding of inverse depth SLAM using a monocular camera. In *Robotics and Automation, 2007 IEEE International Conference on*, pages 2797–2802. IEEE.

[Ramezani et al., 2017] Ramezani, M., Khoshelham, K., and Kneip, L. (2017). Omnidirectional visual-inertial odometry using multi-state constraint Kalman filter. In *Intelligent Robots and Systems (IROS), 2017 IEEE/RSJ International Conference on*, pages 1317–1323. IEEE.

[Ramisa et al., 2009] Ramisa, A., Tapus, A., Aldavert, D., Toledo, R., and Lopez de Mantaras, R. (2009). Robust vision-based robot localization using combinations of local feature region detectors. *Autonomous Robots*, 27(4):373–385.

[Rohatgi and Saleh, 2015] Rohatgi, V. K. and Saleh, A. M. E. (2015). *An introduction to probability and statistics*. John Wiley & Sons.

[Ross, 2017] Ross, P. E. (2017). CES 2017: Nvidia and Audi say they'll field a level 4 autonomous car in three years. In *Institute of Electrical and Electronics Engineers*, volume 5.

[Sandberg et al., 2010a] Sandberg, H., Teixeira, A., and Johansson, K. H. (2010a). On security indices for state estimators in power networks. In *First Workshop on Secure Control Systems (SCS), Stockholm*.

[Sandberg et al., 2010b] Sandberg, H., Teixeira, A., and Johansson, K. H. (2010b). On security indices for state estimators in power networks. In *First Workshop on Secure Control Systems*.

[Santoso et al., 2017] Santoso, F., Garratt, M. A., and Anavatti, S. G. (2017). Visual–inertial navigation systems for aerial robotics: Sensor fusion and technology. *IEEE Transactions on Automation Science and Engineering*, 14(1):260–275.

[Saurer et al., 2017] Saurer, O., Vasseur, P., Boutteau, R., Demonceaux, C., Pollefeys, M., and Fraundorfer, F. (2017). Homography based egomotion estimation with a common direction. *IEEE Transactions on Pattern Analysis and Machine Intelligence*, 39(2):327–341.

[Scaramuzza and Fraundorfer, 2011] Scaramuzza, D. and Fraundorfer, F. (2011). Visual odometry [tutorial]. *IEEE Robotics Automation Magazine*, 18(4):80–92.

[Se et al., 2005] Se, S., Lowe, D. G., and Little, J. J. (2005). Vision-based global localization and mapping for mobile robots. *IEEE Transactions on robotics*, 21(3):364–375.

[Senlet and Elgammal, 2012] Senlet, T. and Elgammal, A. (2012). Satellite image based precise robot localization on sidewalks. In *Robotics and Automation (ICRA), 2012 IEEE International Conference on*, pages 2647–2653. IEEE.

[Shariatmadari et al., 2015] Shariatmadari, H., Ratasuk, R., Iraji, S., Laya, A., Taleb, T., Jäntti, R., and Ghosh, A. (2015). Machine-type communications: current status and future perspectives toward 5g systems. *IEEE Communications Magazine*, 53(9):10–17.

[Shen et al., 2020] Shen, B., Wang, Z., Wang, D., and Li, Q. (2020). State-saturated recursive filter design for stochastic time-varying nonlinear complex networks under deception attacks. *IEEE Transactions on Neural Networks and Learning Systems*, 31(10):3788–3800.

[Shen et al., 2017] Shen, M., Zhao, D., Sun, J., and Peng, H. (2017). Improving localization accuracy in connected vehicle networks using rao-blackwellized particle filters: Theory, simulations, and experiments. *arXiv preprint arXiv:1702.05792*.

[Shen et al., 2014] Shen, S., Mulgaonkar, Y., Michael, N., and Kumar, V. (2014). Multi-sensor fusion for robust autonomous flight in indoor and outdoor environments with a rotorcraft MAV. In *Robotics and Automation (ICRA), 2014 IEEE International Conference on*, pages 4974–4981. IEEE.

[Shi et al., 2021] Shi, L., Liu, Q., Shao, J., and Cheng, Y. (2021). Distributed localization in wireless sensor networks under denial-of-service attacks. *IEEE Control Systems Letters*, 5(2):493–498.

[Shoukry et al., 2017a] Shoukry, Y., Nuzzo, P., Puggelli, A., Sangiovanni-Vincentelli, A. L., Seshia, S. A., and Tabuada, P. (2017a). Secure state estimation for cyber-physical systems under sensor attacks: A satisfiability modulo theory approach. *IEEE Transactions on Automatic Control*, 62(10):4917–4932.

[Shoukry et al., 2017b] Shoukry, Y., Nuzzo, P., Puggelli, A., Sangiovanni-Vincentelli, A. L., Seshia, S. A., and Tabuada, P. (2017b). Secure state estimation for cyber-physical systems under sensor attacks: A satisfiability modulo theory approach. *IEEE Transactions on Automatic Control*, 62(10):4917–4932.

[Shuster and Oh, 1981] Shuster, M. D. and Oh, S. D. (1981). Three-axis attitude determination from vector observations. *Journal of Guidance, Control, and Dynamics*, 4(1):70–77.

[Simon, 2010] Simon, D. (2010). Kalman filtering with state constraints: a survey of linear and nonlinear algorithms. *IET Control Theory & Applications*, 4(8):1303–1318.

[Simon and Chia, 2002] Simon, D. and Chia, T. L. (2002). Kalman filtering with state equality constraints. *IEEE Transactions on Aerospace and Electronic Systems*, 38(1):128–136.

[Smith and Singh, 2006] Smith, D. and Singh, S. (2006). Approaches to multisensor data fusion in target tracking: A survey. *Knowledge and Data Engineering, IEEE Transactions on*, 18(12):1696–1710.

[Song and Chandraker, 2014] Song, S. and Chandraker, M. (2014). Robust scale estimation in real-time monocular sfm for autonomous driving. In *Proceedings of the IEEE Conference on Computer Vision and Pattern Recognition*, pages 1566–1573.

[Strasdat et al., 2010] Strasdat, H., Montiel, J., and Davison, A. J. (2010). Scale drift-aware large scale monocular SLAM. *Robotics: Science and Systems VI*.

[Strohmeier et al., 2018] Strohmeier, M., Walter, T., Rothe, J., and Montenegro, S. (2018). Ultra-wideband based pose estimation for small unmanned aerial vehicles. *IEEE Access*, 6:57526–57535.

[Suarez et al., 2016] Suarez, A., Heredia, G., and Ollero, A. (2016). Cooperative sensor fault recovery in multi-UAV systems. In *Robotics and Automation (ICRA), 2016 IEEE International Conference on*, pages 1188–1193. IEEE.

[Suh, 2018] Suh, Y. S. (2018). Computationally efficient pitch and roll estimation using a unit direction vector. *IEEE Transactions on Instrumentation and Measurement*, 67(2):459–465.

[Sun et al., 2015] Sun, X., Cai, C., and Shen, X. (2015). A new cloud model based human-machine cooperative path planning method. *Journal of Intelligent & Robotic Systems*, 79(1):3–19.

[Tanner et al., 2003] Tanner, H., Jadbabaie, A., and Pappas, G. J. (2003). Coordination of multiple autonomous vehicles. In *IEEE Mediterranean Conference on Control and Automation*, pages 869–876.

[Teixeira et al., 2018] Teixeira, L., Maffra, F., Moos, M., and Chli, M. (2018). VI-RPE: Visual-inertial relative pose estimation for aerial vehicles. *IEEE Robotics and Automation Letters.*

[Theodor and Shaked, 1996] Theodor, Y. and Shaked, U. (1996). Robust discrete-time minimum-variance filtering. *IEEE Transactions on Signal Processing*, 44(2):181–189.

[Tibshirani, 1996] Tibshirani, R. (1996). Regression shrinkage and selection via the lasso. *Journal of the Royal Statistical Society. Series B (Methodological)*, pages 267–288.

[Tong et al., 2018] Tong, X., Li, Z., Han, G., Liu, N., Su, Y., Ning, J., and Yang, F. (2018). Adaptive EKF based on HMM recognizer for attitude estimation using MEMS MARG sensors. *IEEE Sensors Journal*, 18(8):3299–3310.

[Urmson and Whittaker, 2008] Urmson, C. and Whittaker, W. R. (2008). Self-driving cars and the urban challenge. *IEEE Intelligent Systems*, 23(2):66–68.

[Usenko et al., 2016] Usenko, V., Engel, J., Stückler, J., and Cremers, D. (2016). Direct visual-inertial odometry with stereo cameras. In *Robotics and Automation (ICRA), 2016 IEEE International Conference on*, pages 1885–1892. IEEE.

[Van Der Merwe et al., 2001] Van Der Merwe, R., Doucet, A., De Freitas, N., and Wan, E. A. (2001). The unscented particle filter. In *Advances in neural information processing systems*, pages 584–590.

[Wahba, 1965] Wahba, G. (1965). A least squares estimate of satellite attitude. *SIAM review*, 7(3):409–409.

[Wang et al., 2021] Wang, C., Wang, Z., Zhang, L., Cao, D., and Dorrell, D. (2021). A vehicle rollover evaluation system based on enabling state and parameter estimation. *IEEE Transactions on Vehicular Technology*, 17(6):4003–4013.

[Wang et al., 2018a] Wang, H., Jiang, R., Zhang, H., and Ge, S. S. (2018a). Heading reference-assisted pose estimation for ground vehicles. *IEEE Transactions on Automation Science and Engineering*, pages 1–11.

[Wang et al., 2018b] Wang, H., Jiang, R., Zhang, H., and Ge, S. S. (2018b). Heading reference-assisted pose estimation for ground vehicles. *IEEE Transactions on Automation Science and Engineering*, pages 1–11.

[Wang et al., 2015] Wang, J.-q., Wang, P., Wang, J., Zhang, H.-y., and Chen, X.-h. (2015). Atanassov's interval-valued intuitionistic linguistic multicriteria group decision-making method based on the trapezium cloud model. *IEEE Transactions on Fuzzy Systems*, 23(3):542–554.

[Wang et al., 2014] Wang, K., Liu, Y.-H., and Li, L. (2014). A simple and parallel algorithm for real-time robot localization by fusing monocular vision and odometry/AHRS sensors. *IEEE/ASME Transactions on Mechatronics*, 19(4):1447–1457.

[Wang and Chirikjian, 2006] Wang, Y. and Chirikjian, G. S. (2006). Error propagation on the Euclidean group with applications to manipulator kinematics. *IEEE Transactions on Robotics*, 22(4):591–602.

[Wang and Chirikjian, 2008] Wang, Y. and Chirikjian, G. S. (2008). Nonparametric second-order theory of error propagation on motion groups. *The International Journal of Robotics Research*, 27(11-12):1258–1273.

[Wei et al., 2013] Wei, L., Cappelle, C., and Ruichek, Y. (2013). Camera/laser/gps fusion method for vehicle positioning under extended nis-based sensor validation. *Instrumentation and Measurement, IEEE Transactions on*, 62(11):3110–3122.

[Wu et al., 2016] Wu, J., Zhou, Z., Chen, J., Fourati, H., and Li, R. (2016). Fast complementary filter for attitude estimation using low-cost MARG sensors. *IEEE Sensors Journal*, 16(18):6997–7007.

[Wu et al., 2013] Wu, Z., Yao, M., Ma, H., Jia, W., and Tian, F. (2013). Low-cost antenna attitude estimation by fusing inertial sensing and two-antenna GPS for vehicle-mounted satcom-on-the-move. *IEEE Transactions on Vehicular Technology*, 62(3):1084–1096.

[Xie et al., 2011] Xie, L., Mo, Y., and Sinopoli, B. (2011). Integrity data attacks in power market operations. *IEEE Trans. Smart Grid*, 2(4):659–666.

[Yang et al., 2017a] Yang, J., Ma, J., Liu, X., Qi, L., Wang, Z., Zhuang, Y., and Shi, L. (2017a). A height constrained adaptive Kalman filtering based on climbing motion model for GNSS positioning. *IEEE Sensors Journal*, 17(21):7105–7113.

[Yang et al., 2017b] Yang, S., Jiang, R., Wang, H., and Ge, S. S. (2017b). Road constrained monocular visual localization using gaussian-gaussian cloud model. *IEEE Transactions on Intelligent Transportation Systems*, 18(12):3449–3456.

[Yang and Shen, 2017] Yang, Z. and Shen, S. (2017). Monocular visual–inertial state estimation with online initialization and camera–imu extrinsic calibration. *IEEE Transactions on Automation Science and Engineering*, 14(1):39–51.

[Ye et al., 2015] Ye, C., Hong, S., and Tamjidi, A. (2015). 6-DOF pose estimation of a robotic navigation aid by tracking visual and geometric features. *IEEE Transactions on Automation Science and Engineering*, 12(4):1169–1180.

[Yun et al., 2008] Yun, X., Bachmann, E. R., and McGhee, R. B. (2008). A simplified quaternion-based algorithm for orientation estimation from earth gravity and magnetic field measurements. *IEEE Transactions on instrumentation and measurement*, 57(3):638–650.

[Zhang and Singh, 2015] Zhang, J. and Singh, S. (2015). Visual-lidar odometry and mapping: Low-drift, robust, and fast. In *Robotics and Automation (ICRA), 2015 IEEE International Conference on*, pages 2174–2181. IEEE.

Index

Multimodal Perception and Secure State Estimation for Robotic Mobility Platforms, First Edition.
Xinghua Liu, Rui Jiang, Badong Chen, and Shuzhi Sam Ge.

Printed and bound by CPI Group (UK) Ltd, Croydon, CR0 4YY
06/09/2022
03146059-0001